PUMPING STATIONS FOR WATER AND SEWAGE

PUMPING STATIONS FOR WATER AND SEWAGE

RONALD E. BARTLETT
F.I.C.E., F.I.P.H.E., F.I.W.E., M.Inst.W.P.C.
Consulting Civil Engineer

APPLIED SCIENCE PUBLISHERS LTD
LONDON

APPLIED SCIENCE PUBLISHERS LTD
RIPPLE ROAD, BARKING, ESSEX, ENGLAND

ISBN: 0 85334 577 5

WITH 27 ILLUSTRATIONS AND 18 TABLES

© APPLIED SCIENCE PUBLISHERS LTD 1974

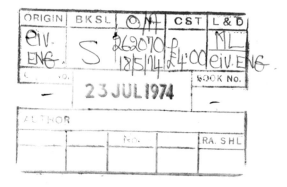

All rights reserved. No part of this publication may be reproduced, stored in a retrieval system, or transmitted in any form or by any means, electronic, mechanical, photocopying, recording, or otherwise, without the prior written permission of the publishers, Applied Science Publishers Ltd, Ripple Road, Barking, Essex, England

Printed in Great Britain by Galliard (Printers) Ltd Great Yarmouth

Preface

'Pumps' have been used to raise water from wells and rivers since Biblical times. These have been developed through the 'lift and force' pumps of the seventeenth century to the piston and centrifugal pumps first introduced in the nineteenth century.

While there are a number of very useful books on the mechanical design of pumps, published information on the civil engineer's side of pumping station design is very limited. Most of the recent technical papers relate to water supply projects and there is very little literature on the design of sewage pumping installations. This general lack of technical information has forced engineers to lean heavily on the pump manufacturers for advice, or to rely almost entirely on their own experience.

This appears to be an appropriate time to summarise present-day practice on pumping station design, particularly as it relates to sewerage and water supply. The emphasis, in the book, on sewage pumping stations is intentional, as is the occasional repetition between various chapters; the latter was to some extent inevitable if the chapters were to be self-contained.

Chapters are included on the various types of pumps, motors and switchgear, together with information on the layout of the buildings and the design of the rising mains. The later chapters cover some aspects of installation, operation and maintenance. For a modern pumping installation (particularly one that is automatically controlled), these various aspects of design cannot be considered in isolation, but must be considered *together* in terms of the complete pumping installation.

The metric (S.I.) system has been used throughout and a conversion table has been included in Appendix C. Nothing has been included on actual capital or running costs as these are so easily out-dated; it is suggested that the reader refer to specialised books and articles for information on costs.

I am indebted to the various manufacturers who have assisted in the provision of plates and figures to illustrate the text, and to my wife for her encouragement and considerable assistance while preparing the manuscript.

R.E.B.

Ashby-de-la-Zouch
Leicestershire

Contents

Preface   v

1. Introduction   1
2. Types of Pump   3
3. Water Supply and Land Drainage   23
4. Sewage Pumping Stations   31
5. Some Special Installations for Sewage, etc.   45
6. Prime Movers   56
7. Pump Starters, Controls and Other Accessories   63
8. Ancillary Equipment   81
9. Buildings   96
10. Pumping Mains   107
11. Specification and Testing   124
12. Operation and Maintenance   129

References and Bibliography   134

Appendices:
A Definitions and Abbreviations   137
B Conversion of Flow Rates   140
C Conversion Factors   142

Index   145

CHAPTER 1

Introduction

For water supply, the decision to use pumps is basic, as it depends on the source of the supply. For wells and boreholes, and for a river intake there is no alternative to pumping. The site of the pumping station will, however, often be governed by other factors than the most efficient plant, and will depend on the source of supply, means of access, and the availability of power supply. The sites of booster pumps and repumping stations will be governed by levels and pressures, and often these pumping stations must be sited in urban areas.

The ideal site for *treating* sewage will be at or near the lowest point of the sewerage system, so that all sewage will flow to it by gravity. This is, however, rarely possible, and there will normally be isolated low-lying areas from which the sewage must be pumped into the main sewers. In flat districts, it may be economical to pump the majority or all of the sewage flow as an alternative to the construction of deep gravity sewers. Where possible, pumping sewage at the treatment works, *after* preliminary treatment, is preferable to the pumping of crude sewage, as the efficiency of the plant will be increased and the capital and maintenance costs will be less. For a sewerage system, the decision to install pumps will often be based on an appraisal of a number of possibilities. From a mechanical point of view, it is preferable to *avoid* the pumping of sewage, or at least, to reduce the number of stations to a minimum.

Various types of pump are available. These include centrifugal, turbine, axial- or mixed-flow, and reciprocating pumps. The choice of the type of pump will generally depend on its duty, in terms of output and head, and also whether for clean water, crude sewage, or something intermediate. All 'rotating element' pumps can have either vertical or horizontal spindles. The choice between vertical and horizontal spindle is generally based on the priming arrangement, and on site conditions; at a site subject to flooding the motors and other electrical gear must be above flood level, or the motors can be of the submersible type.

The development of fully-submersible electric motors has opened up new areas in pumping station design for both sewerage and water supply. Submersible sewage pumps are now quite normal, particularly for the smaller installations. For water supply, submersible motors are used both with well and borehole pumps, and for boosting.

The high standard of modern pumping plant, and the development of automatic control and (latterly) of telemetry, has meant that both water and sewage pumping stations can be operated unmanned, with only routine maintenance. The design of an automatic sewage pumping station is complicated by the very nature of the liquid, and by the variations in the rate of flow throughout the day. There is a very definite relationship between the design of a sewage pumping station and the design of both the sewerage system and the treatment works.

In the design of pumping stations, the civil engineer does not need to have a detailed mechanical knowledge of pump design, but he must have sufficient information on the various types of pump, motor, starter, etc. to be able to choose the most suitable plant and to be able to install it to its best advantage.

CHAPTER 2

Types of Pump

Pumps used for water supply and sewerage can be divided into three broad categories:
(i) those employing rotating elements,
(ii) reciprocating pumps, and
(iii) those using air or steam (*i.e.* pneumatic).

Rotating element pumps include volute centrifugal pumps, turbine, axial-flow, and mixed-flow pumps. Also included in this category would be the rotary pump, the screw pump, and the helical rotor pump. The rotary pump has been popular as a gear-type pump since the 17th century and has been developed more recently as a glandless pump suitable for very small flows. All rotating element pumps impart energy to the water by the application of torque from a shaft (or shafts) driven by a prime mover.

Reciprocating pumps range from the old lift and force 'village pump' type to the modern 'force pump' used in water supply installations. Reciprocating pumps may be used in wells and boreholes (although multi-stage centrifugal pumps are now more common) and they have been popular for many years for pumping sewage sludge.

Under the term 'pneumatic pumps' we can include those employing free air (air lift pumping) and also the various forms of compressed air ejectors. The hydraulic ram might also be considered in this category as it uses a low pressure water supply to compress air, which in turn provides the power for lifting a proportion of the water.

VOLUTE CENTRIFUGAL PUMPS

First developed in the 18th and 19th centuries, these pumps are now used for a variety of duties including both water supply and sewerage.

They are often referred to simply as *centrifugal* pumps since the head is developed principally by centrifugal force. The inlet to the pump is axial and the outlet is tangential, the flow being converted to radial flow as it passes round the *volute* or spiral chamber surrounding the impeller blades. The increase in the cross-section of the volute produces the change from velocity head to pressure head; the volute centrifugal pump has no diffuser passages (see below) and is therefore suited to pumping large volumes against relatively low heads and at low running speeds.

Split-casing pumps are frequently used, as these allow the moving parts to be inspected; the shaft (complete with impeller), together with neck rings and bearings, can be removed without dismantling any pipe joints.

These pumps have impellers made up of a series of curved blades radiating outwards from a central eye. When the impellers are designed for clean water duties the efficiency of the pump can be as high as 90%; these high efficiency units can also be used for pumping treated sewage effluent or (after modification) for sewage which has been passed through fine screens. The *approximate* outputs of centrifugal pumps designed for clear cold water are given in Table 2.1.

TABLE 2.1

APPROXIMATE OUTPUTS OF CENTRIFUGAL PUMPS
(CLEAN WATER)

Diameter of suction branch (*mm*)	Discharge (m^3/h)
100	50
150	200
300	800
600	3 000
900	7 000

The various 'unchokable' and 'disintegrator' pumps are centrifugal pumps with specially designed impellers; these will often have efficiencies of about 50 to 60%. To minimise wear and maintenance costs, the speed of rotation of a sewage pump should be comparatively low; generally not higher than 960 rev/min. Pumps designed to pass solids may have shrouded impellers with the passage between the shrouds capable of passing solids equal to or larger than the bore of

2 TYPES OF PUMP

the suction. The more satisfactory unchokable pump has only one shroud; the impeller then has no 'eye' but generally consists of only one or two S-shaped blades which run completely across the impeller, the blades being secured to the shroud or baseplate. The absence of the eye and of the leading edges of the impellers reduces the possibility of choking with rags and other fibrous material. The absence of the second shroud generally also reduces blockage by rags and it minimises the probability of grit settling between the shrouds and the volute casing. An unchokable pump will be classified according to the size of solids to be passed; the maximum is normally specified as 100 mm diameter. A *multi-stage* unchokable pump is not a practical proposition, but when necessary several pumping units can be connected together in series.

When a centrifugal pump is mounted horizontally, the delivery can be arranged to be either vertical or horizontal, or at any intermediate angle. When the spindle is vertical, as is frequently the case for sewage pumping duties, adequate thrust bearings must be fitted. A centrifugal pump will not begin to lift until the volute chamber is full of water; it must, therefore, either be arranged to operate below the level of the liquid in the suction sump, or some method of priming must be incorporated. Centrifugal pumps can operate against a closed valve for a considerable time without damage to either pump or motor; damage can, however, be caused due to overheating.

TURBINE PUMPS

The term *turbine pump* is usually given to the type of multi-stage centrifugal pump often used in water supply engineering. These pumps are frequently used as horizontal spindle units; they also form the basis of some of the centrifugal borehole and submersible pumps. In the turbine pump, the velocity head induced by the moving impellers is transformed into a pressure head by means of diffuser chambers, before the water is passed to the impeller of the next stage. The diffuser chambers are fitted with tapering guide passages discharging into spiral passages. In this manner, comparatively small quantities of water can be lifted through quite high heads.

The impellers of turbine pumps are usually of bronze and are often the double-shrouded type. The design of the impellers and guide passages makes this type of pump quite unsuitable for use with sewage or other liquids containing solid matter. Should the head on the pump be reduced below its design level, the motor can be overloaded; for this reason, many multi-stage borehole pumps are fitted with mixed-flow impellers (see below) which have non-overloading characteristics over a wide variation of head.

AXIAL-FLOW PUMPS

Axial-flow, or propeller, pumps which discharge axially without a volute chamber have been developed during this century. The impeller is similar in design to the screw of a ship; this operates in a 'chamber' which in fact may be of the same diameter as the suction

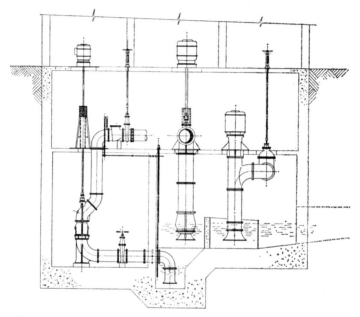

Fig. 2.1. *Axial and mixed-flow pumps* (by courtesy of Sigmund Pulsometer Pumps Ltd.).

and delivery pipes; the delivery pipe may then be bent through 90° to provide access for the drive shaft as shown in Fig. 2.1. In some respects the axial-flow pump is similar in its action to the Archimedian screw (see below).

Axial-flow pumps can be vertical, horizontal, or inclined; they are frequently used for land drainage work and for water supply installations where large quantities have to be pumped against comparatively low heads (up to 14 or 15 m). With suitable modifications they can be used for handling storm sewage and also for recirculation or effluent pumping at sewage treatment works. Their efficiency can be up to 75 to 90%, while their relatively high rotative speeds result in lower motor costs.

As the impellers of axial-flow pumps are rarely fitted with shrouds they can more easily be designed for variation in pitch of the blades. When required, this variation in pitch, to adjust the output of the pump, can be carried out while the pump is in operation.

MIXED-FLOW PUMPS

These pumps are, in effect, intermediate between volute centrifugal pumps and axial-flow pumps, although for general classification they are frequently classed along with axial-flow pumps, as the two types are used for similar duties. The delivery head is developed partly by centrifugal force and partly by lift from the impeller vanes. Flow enters the pump axially and discharges in an axial and radial direction, normally into a volute type of casing.

Mixed-flow pumps are suitable for moderate heads (up to about 25 to 35 m), and are more usually used for water supply pumping than for sewerage. Mixed-flow units are employed in many shaft-driven borehole pumps and in the larger submersible pumps; they are non-overloading under open valve conditions.

ROTARY PUMPS

Gear-type pumps have been used for many years as positive displacement pumps and they are still used in the lubricating systems of diesel and gas engines. Their capacities are quite small and their efficiencies are generally around 40%.

Various types of rotary and semi-rotary pump have been developed for water supply; these have found applications in private installations where very small quantities are required. One of these pumps, suitable for flows up to about 1·0 to 1·5 litres/sec (3·5 to 5·5 m^3/h), is illustrated in Fig. 2.2. This type of rotary glandless pump can also be used for lifting sewage from isolated properties through a total static lift of up to about 10 m.

SCREW PUMPS

The screw pump is an extended version of the axial-flow, or propeller, pump. It is based on the principle of the Archimedes screw and has

Fig. 2.2. A rotary pump (by courtesy of Tuke and Bell Ltd.).

been developed during recent years for pumping sewage (particularly at a sewage treatment works), and for land drainage pumping. Screw pumps are available with outputs up to 25 000 m^3/h against heads of up to about 10 m; the maximum head is dependent on the length of the screw. Pump efficiencies are quoted as between 65 and 75%. The slow action (between 20 and 90 rev/min) makes this type of pump very suitable for pumping activated sludge or other liquids containing flocs. As no wet well is required the cost of the installation can be kept to a minimum. A screw pump installation is shown in Fig. 2.3.

HELICAL ROTOR PUMPS

In this type of screw pump, a single helical rotor revolves within a double internal helical stator, maintaining a constant seal across the stator, and giving a uniform positive displacement. The principle of operation is illustrated in Fig. 2.4. Capacities of up to 75 m^3/h are possible. This type of pump is suitable for sewage and for other liquids containing materials in suspension; it can also be used as a borehole pump for domestic water supply.

2 TYPES OF PUMP

Fig. 2.3. *An installation of screw pumps* (by courtesy of New Haden Pumps Ltd.).

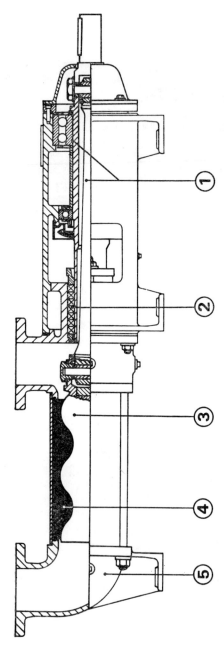

Fig. 2.4. The helical rotor pump. 1. Shaft and universal drive. 2. Only one gland. 3. Rotors. 4. Stator of resilient rubber or other material. 5. End cover and barrel removable for inspection (by courtesy of Mono Pumps Ltd.).

RAM (LIFT AND FORCE) PUMPS

In the ram type of pump a plunger moves to and fro (*i.e.* it reciprocates) inside the pump, alternately lifting liquid (on the suction stroke) and delivering it (on the delivery stroke). In a double-acting pump, each side of the pump is fitted with both inlet and delivery valves so that the liquid is delivered into the pumping main more-or-less continuously when the pump is operating. There will however be large variations in flow throughout the full cycle of the stroke and to even out the flow, waterworks installations are almost invariably fitted with air vessels on both suction and delivery; three-throw or similar type pumps are generally used.

Ram pumps are suitable for high heads and they usually operate at low speeds; the *average* speed of the piston is often not more than 1·0

Fig. 2.5. *A plunger pump for use with sludge* (by courtesy of **Pegson** Ltd.).

to 1·25 m/sec. While force pumps are still in use in waterworks practice (particularly in boreholes) they are tending to be replaced by centrifugal pumps; ram pumps were particularly suitable as steam-driven direct-acting pumps. The use of plunger pumps in sewage pumping is now generally limited to the smaller installations; plunger pumps suitable for crude sewage (after screening) are available for flows up to about 15 or 20 litres/sec (*see* Fig. 2.5). Plunger pumps are frequently used for pumping sewage sludge where the quantities do not exceed about 15 litres/sec.

Reciprocating pumps have high *pump* efficiencies (up to 95%) and they retain their efficiency through large variations in head conditions. As the power requirement varies almost directly with the head, this type of pump is very suitable for conditions where the operating head will vary considerably.

DIAPHRAGM PUMPS

The diaphragm pump is a special type of reciprocating (or ram) pump, and is operated by the buckling of a flexible metal or rubber diaphragm (or disc) when moved by a crank or lever connected to its centre. Non-return valves are fitted at the inlet and outlet and the movement of the disc alternately sucks up liquid into the pump chamber and discharges it to the rising main. Diaphragm pumps can be either single- or double-acting. Some types of pump are fitted with air chambers on the suction and delivery mains to even out flows and to give larger outputs at low heads.

The main use of this type of pump is as a contractor's pump for clearing water from shallow wells, trenches, etc. as it is particularly suitable for lifting muddy water; specially designed models are available for pumping sewage and other sludges. While the smaller pumps are hand-operated, it is more normal to incorporate a motor drive; this is frequently a petrol engine, but pumps with either diesel engines or electric motors are also available.

PNEUMATIC EJECTORS

It is not possible to manufacture a normal centrifugal pump capable of passing very low flows and at the same time to make it suitable for liquids containing solids up to 100 mm diameter. Various special devices have been manufactured for such conditions, including ejectors operated with compressed air.

2 TYPES OF PUMP

A pneumatic ejector consists of a cast iron vessel into which the liquid (usually sewage) either flows by gravity or into which it is drawn by suction. When the body of the ejector is full, the contents are *ejected* into the rising main by the action of the compressed air. The automatic operation is controlled by floats and a system of reflux valves. Compressed air is supplied by a separate automatic compressor which can be used to charge one or more air containers.

Ejectors can either be gravity filled and installed just below the sewer invert (*see* Fig. 2.6) or of the 'lift and force' type installed at

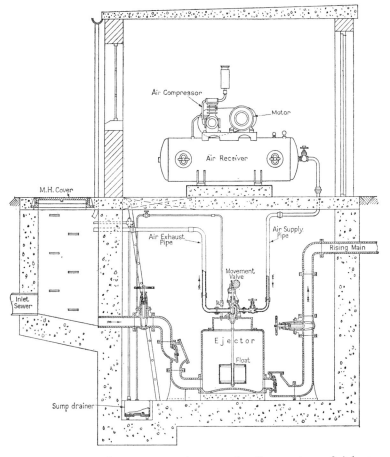

Fig. 2.6. A typical gravity-type ejector station (by courtesy of Adams-Hydraulics Ltd.).

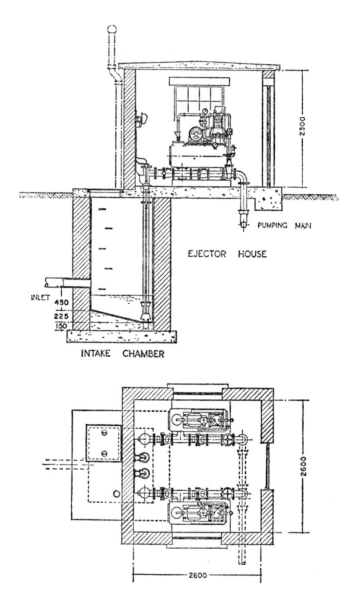

Fig. 2.7. A lift and force type of ejector (by courtesy of Tuke and Bell Ltd.).

ground level (Fig. 2.7). Ejectors have low efficiencies (between 20 and 50%) and are comparatively expensive in capital cost as compared with small pumps. They are, however, very reliable, they require little maintenance, and they are capable of dealing with small flows of sewage containing solids without the necessity of either a suction well or a screen. Other advantages include the complete enclosure of the sewage (so that there should be no problem of smell), and fully automatic operation. Ejectors are most suitable for moderate heads (not more than about 30 m) and for flows from about 3 m^3/h (0·001 cumec) to about 30 m^3/h. They are therefore suitable for dealing with the sewage from isolated properties or from groups of up to 100 or more houses. Ejectors are usually installed in duplicate so that one is filling while the other is discharging; a duplicate compressor is not essential but should be provided if a breakdown would have serious results. Although it is possible for a number of ejector installations to be supplied from one central air compressing station, this is not very common and it is now normal for each ejector station to be self-contained.

Ejectors are sometimes used at a sewage treatment works to lift sludge from isolated 'desludging' points to a dewatering plant.

AIR LIFT PUMPS

Air lift pumps are used in some smaller water supply boreholes, either for temporary pumping duties while the borehole is being test-pumped or, occasionally, in the permanent pumping installation. At a sewage treatment works, air lift pumping is useful for emptying tanks where screening is not necessary and where the additional aeration of the sewage will assist in purification.

An air lift pumping installation consists of two vertical pipes—a small diameter pipe to supply the air and a larger one to act as the rising main (*see* Fig. 2.8). As the air leaves at the bottom of the air pipe it expands and lowers the density of the water or sewage in the lower part of the rising main; the pressure of the surrounding liquid then forces this water upwards in the main. As the system has no moving parts, its efficiency is not impaired by grit, etc. and it is very good for use with sludges and dirty liquids.

The lift that is possible is dependent on the depth of submergence of the air pipe (the ratio of the submergence to the lift) and on the quantity of air supplied. A formula and calculations for these are given in Chapter 6. Efficiencies are low and are rarely higher than

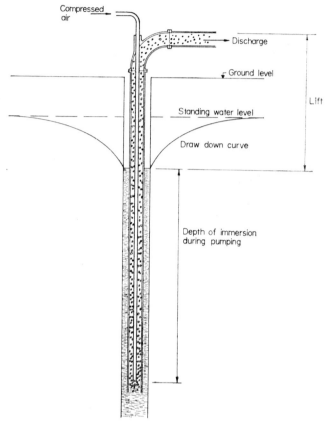

Fig. 2.8. Air-lift pumping.

about 50%. One compressor may sometimes be used to supply more than one pumping unit.

THE PNEU PUMP
This small pump is available in stock sizes with outputs up to 3 litres/sec. It has no mechanical components or glands and relies for its operation on a simple ball valve at the inlet pipe, together with an air seal on the delivery. Air is supplied from an external compressor, and the supply of air (to bring the pump into operation) can be controlled by a simple float switch. The pneu pump is claimed to be suitable for most liquids and slurries through a large temperature range and is therefore very useful as a sump pump.

2 TYPES OF PUMP

HYDRAULIC RAMS

In the hydraulic ram, the pressure obtained by suddenly stopping a moving column of water is used to lift part of that water to a higher level. No external power supply is required, and no lubrication is necessary. The moving flow of water is used to close a valve on the downstream side; the pressure then builds up in the supply main sufficiently to force some of the water past a valve into an air chamber. As the column of water is brought to rest by the air chamber the recoil closes the entry valve to that chamber; this allows a reduction in pressure which allows the outlet valve to re-open to complete the cycle. The inlet and outlet valves are adjustable to regulate the flow of water, while a 'snifting valve' replenishes the air in the air chamber.

Hydraulic rams can be used for individual water supplies to buildings where there is a suitable source of supply some distance below them. They are available for flows up to over 100 litres/sec; the supply 'fall' is usually limited to about 6 m to prevent excessive wear on the valves. The proportion of the water raised to a higher level is directly related to the ratio of the fall of the main water supply to the ram, and the height to which the supply is to be raised.

It is generally accepted that a hydraulic ram has a mechanical efficiency of about 50%, and can be used to lift about 15% of the water to a height four times the fall to the ram. These figures can be varied pro rata, i.e. $7\frac{1}{2}$% to eight times the fall, 30% to twice the fall, etc. The diameter of the supply pipe is often set at double that of the discharge main. The latter can be calculated approximately from the formula:

$$d = 37\,(Q)^{\frac{1}{2}} \qquad \textbf{Formula 2.1}$$

where

d is the diameter of the discharge main, in mm,

Q is the quantity of water to be supplied, in litres/sec.

As a ram will not operate under water, it must be sited so that it is above maximum flood level; this must be taken into account when calculating the head available. Rams are available which can use impure water as the power supply to raise a supply of pure water from another source. The Adams 'Autoram' uses this principle to take advantage of a flow of high level sewage or river water to create air pressure to raise small quantities of sewage from lower areas.

SPECIFIC SPEED

The choice between a centrifugal, a mixed-flow, or an axial-flow pump is basically a matter of comparison of speed, output and head conditions. With all three types, the flow increases as the head decreases. The decrease in head is more rapid with axial-flow pumps than with centrifugal or mixed-flow pumps.

Pumps are conveniently compared by reference to their 'specific speeds'. *Specific speed* can be defined as that speed, in revolutions per minute, at which an impeller generally similar to the one under consideration, and reduced in size, will develop unit head at unit output. The formula for specific speed is:

$$n_q = \frac{n(Q)^{\frac{1}{2}}}{H^{\frac{3}{4}}}$$ Formula 2.2

where

n_q is the specific speed, in rev/min,
n is the actual speed, in rev/min,
Q is the capacity of the pump, in m³/h,
H is the total manometric head, in metres.

The three types of 'rotating element' pumps have specific speed ranges, in the above terms, as follows:

Centrifugal: up to 5000
Mixed-flow: 5000 to 10 000
Axial-flow: 10 000 to 20 000

Although specific speed is essentially a matter for the pump manufacturer, it is useful for the design engineer to have a general understanding of the subject, as it will give him an indication of the type of pump which will be required for a particular duty.

CHARACTERISTIC CURVES

In general terms, for any specific design of centrifugal pump, the quantity of liquid pumped will vary directly as the speed of rotation; the head varies as the square of the speed; and the power required varies as the cube of the speed. The speed of rotation is chosen by the manufacturer to provide sufficient velocity to overcome the total manometric head. At any particular speed, the delivery of a pump will vary with change of head according to the pump's *characteristic curves*; the output of the pump will decrease as the head increases until it finally becomes nil. Along with variations in head there will be variations in pump efficiency, with the efficiency curve having a maximum at some intermediate position.

There are therefore three curves: (i) head/discharge, (ii) power input/discharge, and (iii) efficiency/discharge (or efficiency/head).

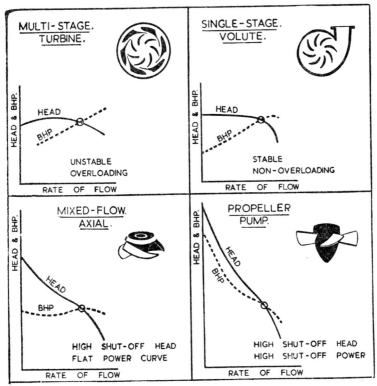

Fig. 2.9. *Four typical pump characteristics* (reproduced from *Manual of British Water Engineering Practice*, 4th Edition, Vol. II, p. 439).

These three curves are plotted together on one sheet of squared paper to form a set of characteristic curves for a pump operating at one chosen speed, *i.e.* they are characteristic of that particular type of pump. The curves for a centrifugal pump will be quite different from those for axial-flow and screw pumps (*see* Fig. 2.9).

The 'design point' for any pump is that point at which the pump operates at maximum efficiency. Fig. 2.10 incorporates a typical set

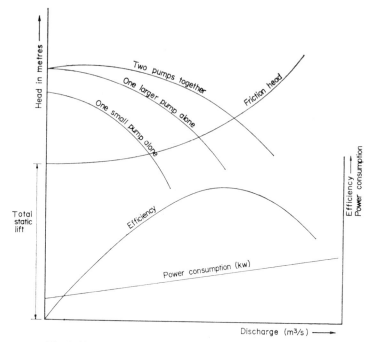

Fig. 2.10. Friction head and pump characteristic curves.

of characteristic curves for a centrifugal pump. Manufacturers often prepare one chart to illustrate the range of performance of a group of pumps running at various speeds; one such chart is reproduced in Fig. 2.11.

CAVITATION

Should the recommended maximum suction lift for a pump be exceeded, there is danger of cavitation, *i.e.* of a partial vacuum being formed so that the pump will be drawing up water containing bubbles of vapour. Axial-flow and mixed-flow type pumps are particularly prone to cavitation. For normal temperate climates, the maximum suction lift will usually be set at between 6·5 and 7·0 m; this should be reduced for liquids of higher temperature (*e.g.* about 6·0 m in tropical countries).

The consequences of cavitation include the erosion of metal surfaces (particularly the impeller blades), audible rattling, vibration, and a lowering in the efficiency of the pump.

2 TYPES OF PUMP

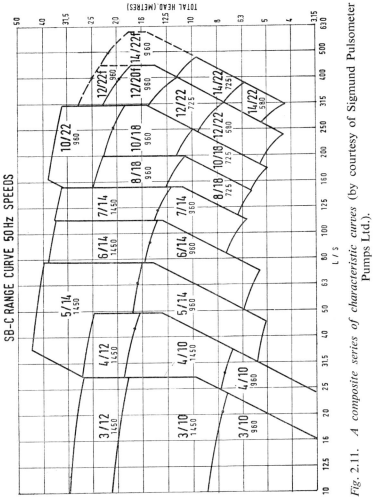

Fig. 2.11. A composite series of characteristic curves (by courtesy of Sigmund Pulsometer Pumps Ltd.).

POWER INPUT

When the output and total manometric head have been calculated, the power input required by the pump can be expressed by the following formula:

$$P = \frac{Q \times H}{3 \cdot 67\, r} \qquad \text{Formula 2.3}$$

where

P is the power input required, in kW,
Q is the output of the pump, in m^3/h,
H is the total manometric head, in metres,
r is the efficiency, expressed as a percentage.

If the total head on a pump of 500 m^3/h output is 21·4 m, and the pump efficiency is 40%, the power requirement of the pump will be:

$$\frac{500 \times 21 \cdot 4}{3 \cdot 67 \times 40} = 73 \text{ kW}$$

To calculate the size of the motor required, further allowance must then be made for the efficiency of the motor. If this is, say, 85% in the above example, then the motor power must be:

$$73 \times \frac{100}{85} = 86 \text{ kW}$$

CHAPTER 3

Water Supply and Land Drainage

Pumps are used frequently in water supply systems, where their duties range from the abstraction of raw water to the boosting of the flow or pressure in a pipeline. Pumping is obviously necessary when the source of water is *below* the area to be supplied; it will also be necessary when there is insufficient head available for the flow by gravity of the quantity required.

The main types of pump in use are:

 (i) Pumping from wells and boreholes
    (a) centrifugal pumps driven by motors located at the surface,
    (b) submersible pumps and motors,
    (c) reciprocating pumps.
 (ii) Pumping from river intakes
    (a) horizontal-spindle centrifugal pumps,
    (b) vertical-spindle centrifugal pumps,
    (c) axial-flow pumps.
 (iii) Pumping after treatment (turbine pumps, etc.).
 (iv) Pumping from one level to a higher level for distribution or storage.
 (v) Boosting to increase the flow, or pressure, in a pipeline.

In waterworks practice, pumping is often the most economical and frequently the only solution to a problem involving the abstraction of the supply of water; it therefore forms a basic part of waterworks engineering. In contrast, in sewerage design (*see* Chapter 4), pumping may only be resorted to after detailed investigation of a number of possible 'gravity' alternatives.

Many waterworks pumps are designed to run continuously for long periods of time; high efficiency is therefore very important, as an increase in efficiency of only 1% can represent a considerable saving in the running costs over a period. The pumps must be reliable in use and they must be designed so that the water being pumped is not contaminated by oil or grease from the pump, or by the by-products of corrosion. These considerations call for a high standard of finish internally. Similarly, the external finish of waterworks pumps is usually of a high standard as this encourages hygiene and cleanliness at the works. Pumps handling cold water are prone to condensation and, therefore, corrosion, and any metal finishes must be suitable for these conditions.

Waterworks pumping plant is usually designed for a specific set of conditions in terms of output and head; any appreciable variation from those conditions will not be expected unless there is some fault in the system, *e.g.* a burst pumping main. While, in the past, considerable use has been made of water or wind power, steam, oil and diesel to drive waterworks pumps, the use of electricity is now much more usual; the supply of electric power is now very reliable and its use permits much more sophisticated methods of control of the pumps.

Pumps installed for land drainage are usually required to lift large quantities of (comparatively) dirty water through quite low heads. These pumps are only required to operate at times of high run-off and they may therefore stand idle for long periods. It is therefore important that their capital cost should be as low as possible and that they are always available for use, even after a period of months standing idle.

WELL AND BOREHOLE PUMPS

The more usual pumps for boreholes and wells will be of the single entry impeller type, either with close-coupled submersible motors, or with the motors at the surface. In some older stations reciprocating pumps are still in use for quite substantial flows; for new schemes, when abstraction is less than about 30 m^3/h, this type of pump may still be more economical than an impeller pump. These pumps may be either single- or double-acting. The rotary motion of the drive motor is converted to reciprocating motion at the top of the borehole, and a shaft then connects to the pump at the bottom; this shaft will be located inside the rising main and will need intermediate supports. Air-lift pumps are also used on some small schemes, particularly overseas where easy maintenance is important, but their efficiency is low.

3 WATER SUPPLY AND LAND DRAINAGE

Well and borehole pumps are usually run more-or-less continuously, and high standards of reliability and efficiency are therefore most important. Most impeller type pumps will be of the multi-stage type, and mixed-flow impellers are general as the diameter is restricted by the size of the borehole. Variable speed drive is possible but rarely relevant, as mixed-flow impellers give non-overloading power characteristics over a wide head variation. Multi-stage radial flow pumps are incorporated in some of the smaller diameter submersible units. When the drive motor is at the surface the shaft will be located *inside* the rising main in the borehole, and intermediate submerged bearings will be provided. Shaft-driven pumps are usually more efficient than submersible units for outputs above about 250 to 300 m^3/h.

With a totally submerged unit, the motor will be located *below* the pump so that the rising main above the pump is unrestricted. The motor will be of the squirrel-cage type and will either have 'wet' windings (the water having access to them), or 'dry' windings as part of a sealed unit. Sealed stators, with the windings separately contained in an oil-filled chamber, are more usual for the smaller type of installation (under about 45 kW). One advantage of the submersible pump over the shaft-driven unit, is its suitability for use in a borehole which is not absolutely straight or vertical. A disadvantage may be a lower motor efficiency (70 to 75%) due to the exaggerated ratio of length to diameter, as compared with a conventional motor (80 to 85% efficiency).

The feed cable to a submersible motor will be taken down on the outside of the rising main along with any signal cable for 'low water' cut-off, when provided. The size of the supply cable will be limited by the facilities for handling the cable at site; single-core cables may be preferable to the use of a three-core cable [18]. The three-phase constant-speed squirrel-cage motors will be started either 'direct-on' or through an auto-transformer; the electricity supply must therefore be suitable for the high starting loads. When two or more submersible pumping units are operated automatically from the same electricity supply, it may be necessary to incorporate a time delay, or other device, to prevent both units starting together.

While the borehole pumps can be arranged to pump to levels above ground level, it is usually preferable to pump to the surface only, and to provide a further set of pumps to raise the water from there. This arrangement will necessitate a break pressure tank (either a closed or open tank) and a suitable system of level controls. When the water is to be treated, this arrangement can be used to provide a point for chemical dosing. In some circumstances, particularly at isolated pumping stations, it may be more convenient to use the borehole pumps for the combined duty, lifting the water from underground and raising it through a limited head to a storage tank above ground level.

RIVER INTAKES

A pumping installation at a river intake will have quite different characteristics from one drawing from a well or a borehole. This type of station will usually be subject to continual variation in head and output as the water is pumped to a reservoir; in addition the duty may be intermittent rather than continuous. Some of the conditions which apply to irrigation and land-drainage pumping installations (see below) are equally applicable to this type of installation.

The quality of the water may be low and any intake will normally include a wet well protected by screens for the removal of larger debris. The water will then be repumped after treatment.

As some temporary stoppage in pumping may be acceptable, the provision of standby equipment is often not vital. The pumps will generally be mixed-flow or axial-flow; in the latter case variable pitch impellers may be used with advantage, particularly if the variation in head is substantial.

Automatic control of the raw water pumps at a river intake will allow the station to operate unattended. Lupton described the Metropolitan Water Board installation at Chingford in a paper in 1956 [13]; that station was equipped with axial-flow pumps with automatically-controlled variable-pitch impellers in the smaller pumps. Control at Chingford was based on the need to keep a residual flow in the river as well as the minimum operating levels in the delivery reservoirs; a further clock-operated control prevented the operation of the larger pumps during electrical peak load periods.

PUMPS FOR IRRIGATION AND LAND DRAINAGE

Pumps for irrigation and land drainage are similar in many ways to those used at river intakes for domestic water supply. The usual requirements for these pumping stations are large quantities and low lifts. These are best met with either axial-flow or mixed-flow pumps. They are available for use with electric motors, diesel or steam, and can be installed vertically, horizontally or inclined. Axial-flow pumps are preferable at very low heads in view of their high specific speeds and lower capital costs. For higher heads mixed-flow pumps are more economical. In the fenlands of eastern England, electrically-driven, vertical-spindle, axial-flow pumps are usually preferred.

By the use of variable speed motors the output can be kept constant over a range of heads, or the output may be varied for a constant head. With axial-flow pumps, variable pitch impellers may be used; these can be controlled manually, or automatic control can be included so that the blade position can be monitored in response to changes in head or output.

The screw pump has also been used for many years in land reclamation schemes and for irrigation. This type of pump becomes a serious competitor to the axial-flow pump when the discharge level remains constant.

Fixed-bar type screens are usually fitted at the intake to prevent the passage of solids which would damage the pumps. If the pumps are to provide water free from debris, *e.g.* circulating water for thermal power stations, it is more usual also to install some form of fine screen; this may be of the drum or disc type. With land drainage pumps of less than about 400 mm diameter, unless particular attention is paid to screening, trouble may be experienced due to weed growth being drawn into the pumps; this can build up and eventually choke the pumps.

Small 'wind mills' mounted on steel towers may be used for irrigation work when comparatively small quantities are involved. A 7·5 m wheel will usually develop sufficient power in a 25 km/h breeze to raise 100 m^3/h to a height of about 4 m. Larger wheels will have correspondingly larger outputs.

HIGH LIFT PUMPS

After pumping from a borehole, or after treatment, it is often necessary to lift the water through a substantial head, either into a service reservoir or directly into supply. These 'high lift' pumps usually have characteristics quite different from the pumps considered so far in this chapter. The heads on these pumps may be as low as about 15 m or they may be up to 100 m or more. Continuity of operation is generally very important, and adequate standby provision must be available. As with borehole plant, the long hours of operation justify the provision of high efficiency plant. Automatic control, based on a water level in the receiving reservoir, is fairly standard.

While occasionally the pumps may be of the mixed-flow type, they are usually of the centrifugal (radial-flow) type. They may be of the single-entry end suction type, of the double suction type, or of the multi-stage turbine type. The multi-stage type will normally comprise a number of single entry impellers on the same drive shaft; the number of stages may vary from two to ten or more. The multi-stage pump can be mounted horizontally, or it can take the form of a suspended unit similar to those used in wells and boreholes. One of the problems with multi-stage pumping is the considerable end thrust which must be restricted by the impeller bearings. Most manufacturers of multi-stage pumps (including multi-stage borehole pumps) have overcome this problem by mounting the impellers back to back (or 50% in one direction and 50% in the opposite direction) to neutralise each other's thrust.

Double suction pumps are manufactured with split casings for easy maintenance, and in units of one, two or three stages. High lift pumping units are frequently equipped with variable speed or multi-speed motors to provide maximum flexibility of control of the output.

BOOSTERS

Booster pumps may be incorporated in a pipeline either to increase the head in the line or to increase its carrying capacity; either of those results can usually be achieved at a much lower cost than would be incurred by duplicating the pipeline. In 1948 [11] Lupton stressed that 'it pays to pump', and he suggested that if the cost of a larger main was compared with that of a smaller main together with a booster for occasional use, 'the pumping installation is often the more economical one'.

Boosters are particularly useful as means of supplementing the supplies to isolated areas where the growth of consumption over a period of years has outgrown the capacity of the supply mains. Boosting is also used where development has taken place in a local high area which cannot be satisfactorily supplied by the existing mains (this situation may also be considered as 'repumping'—see below). The head on the pump will vary from a maximum design head down to zero. The conventional booster installation once comprised horizontal spindle centrifugal pumps in a small pumphouse at ground level. More recently, considerable use has been made of

3 WATER SUPPLY AND LAND DRAINAGE

submersible pumps fitted either horizontally or vertically in a small chamber below ground; these are particularly useful for installation in damp conditions. Submersible units may be fitted within the mains or within shrouds.

Boosters are almost invariably driven by electric motors and, where relevant, these can be automatically controlled for unmanned operation. As the need for boosting is frequently restricted to the hours of peak demand, the pumps may stand idle for some hours each day. Under these conditions mechanical gland seals are generally more satisfactory than conventional soft packings which may tend to dry out while the pumps are idle.

Boosters will nearly always be automatically controlled; their operation may either be based on the rate of flow or the head downstream, or by time controls. Control at predetermined times, or when the pumps discharge to an elevated tank, is fairly straightforward. Where no storage is available it may be convenient to install a pneumatic compression tank; the pump is then started and stopped by means of a diaphragm type pressure-operated switch. The air in the pressure cylinder is replenished by means of a small air compressor.

REPUMPING

This is, to some extent, a form of boosting (see above). A small repumping station can be provided to deal with one or more local high spots, or tall buildings, in a distribution system more economically than a general increase in pressure over a large area. As with boosting, these pumps are usually automatically controlled, either by a float switch in a high level tank, or by a pressure switch. The use of a pneumatic compression tank (see above) is ideal for this type of installation, and is also a safeguard against surge pressures in the delivery pipeline.

RECIPROCATING PUMPS IN WATER SUPPLY

While reciprocating pumps were used in the past on many water supply projects, particularly as well pumps, this type of pump has now been almost completely superseded by the various types of rotodynamic pumps. Some older installations still exist which incorporate reciprocating pumps; their efficiency was often higher than a comparable centrifugal pump, particularly when operating at high heads.

Modern installations of reciprocating pumps are generally restricted to the smaller sizes, and in particular, where there may be considerable variation in the delivery head. As discussed above ('well and borehole pumps') this type of installation is relevant only when the rate of abstraction is less than about 30 m^3/h.

SMALL INSTALLATIONS

A number of single-stage pumping units are available which are suitable for the smaller installations, *e.g.* isolated factories, farms or domestic properties. The smallest of these comprise centrifugal pumps with 25 mm suctions and deliveries, and are available with cast iron casings (and bronze impellers, etc.) or in plastics. These pumps are suitable for suction lifts of up to about 8 m.

Submersible pumps are also available for boreholes of small diameter (minimum 100 mm) and with outputs of less than 1 m^3/h. Air lift pumping is used in small wells and boreholes, both in this country and overseas.

CHAPTER 4

Sewage Pumping Stations

When a sewerage system is being designed, pumping is generally avoided if possible as the pumps and allied equipment automatically form a 'weak point' in the system. In many cases, however, pumping will be the only satisfactory or economical method of lifting the sewage from low-lying areas to the treatment, or outfall, works; this will particularly apply to isolated low-lying areas where it would be uneconomical to lay the main sewer at a low level merely to serve a few properties. A station will normally include at least two pumps so that one is available as stand-by; if the variations in flow are wide, the station may consist of a number of different sizes of pumps so that the installation will operate satisfactorily both at minimum and maximum rates of flow of the incoming sewage.

Every effort should be made to avoid the necessity to pump *storm* sewage in view of the large quantities involved at comparatively infrequent intervals, resulting in high capital costs and high overheads (kVA charges and maintenance). Where foul sewage, or a mixture of foul and storm sewage, is to be lifted, the pumps must be capable of passing any solids likely to be discharged from house drains. The maximum size of these solids is usually taken as 100 mm. An alternative may be to install one of the small macerator pumps now available; for very small installations these have the added advantage of only requiring a small diameter rising main, with a resultant self-cleansing velocity at quite low flow rates.

The siting of a sewage pumping station will be dictated, to a great extent, by the layout of the development which it will serve, and by the details of the sewerage system. As far as possible, precautions must be taken to avoid nuisance from odours or noise, this applies not only to existing development but also to probable future development. Odours are particularly likely to be troublesome when the sewage becomes septic or if screens are incorporated into the station. A sewage pumping station should also be carefully sited so that an overflow is possible in the event of power failure. The external appearance of a station should blend with the development; in some circumstances an underground station may be the best solution. Considerations must also be given to known or probable flood levels in the area.

Pumps installed at a sewage treatment works may be for crude sewage (at the works inlet) or for treated effluent, or for some intermediate stage. The design of the pumps will then depend on the type and amount of solids to be pumped. Pumping is also often necessary for the sludge which has been settled out from the sewage in either the primary or final tanks; this may be 'activated sludge' to be returned to the treatment plant, it may either gravitate or be pumped to tanks on the site prior to treatment, or it may be pumped off the site for treatment elsewhere.

DRY WEATHER FLOW

The dry weather flow (d.w.f.) used in sewerage calculations should take into account the flow of domestic sewage plus any infiltration, together with the flow of industrial wastes. This can be expressed, in litres per day, as:

$$\text{d.w.f.} = PQ + I + E \qquad \textbf{Formula 4.1}$$

where

P is the population served,
Q is the average domestic water consumption, in litres/day,
I is the rate of infiltration, in litres/day,
E is the volume of industrial effluent (in litres) discharged to the sewers in 24 hours.

The rates of flow used are generally those which occur after a period of six or seven consecutive days of dry weather during which the rainfall has not exceeded 0·25 mm.

Water usage per head tends to increase continually, and is now rarely less than 140 litres per day. A *maximum* domestic sewage dry weather flow could be taken as 230 litres per person per day, but higher figures have been used from time to time, especially in the design of sewers for new towns, where every house will have modern plumbing and drainage facilities. It would seem wise to base the

4 SEWAGE PUMPING STATIONS

design of future *urban* schemes on a domestic d.w.f. of 230 litres per person per day. A figure of 150 litres might be more suitable for rural areas.

PUMPING STATION CAPACITIES

A sewage pumping station will usually be designed to have a pumping capacity of up to six times dry weather flow (6 d.w.f.), this being the normal basis of design of the sewers themselves if they are more-or-less 'separate', or if suitable storm sewage overflows are incorporated upstream of the pumps. Where the system is new, and it is entirely on the separate system, there may be some justification for reducing this figure to 4 d.w.f. for the sewers themselves. This may not be advisable, however, as far as pumping station design is concerned, and any reduction in total capacity must be made with caution, full consideration being given to the possibility of infiltration into either the sewers themselves or into any house drainage system which may be connected to the system in due course. Where any infiltration is to be expected, or where the sewerage system is not strictly separate, both the sewers themselves and any pumping stations must be capable of dealing with flows of up to at least 6 d.w.f. Where relevant, of course, additional provision must be made for the flows from farms, etc. over and above the basic domestic and industrial flows.

When the sewerage system is 'combined', flows considerably in excess of six times d.w.f. can be expected in times of storm. If pumps are installed for flows up to 6 d.w.f., provision must be made for the excess flows to be diverted at one or more *storm sewage overflows*. These can be sited on a sewer where a convenient watercourse is available; or it may be convenient to have a storm sewage overflow at the inlet to the pumping station.

If the sewers are combined and an overflow is not possible, the pumps must be capable of dealing with very wide variations in the flow, and special consideration must be given to the maximum capacity to be provided. When they are available, past records of flow will be useful; provision may also have to be made for any future increases in the impermeable area draining to the pumping station.

If the probable development of the area draining to the station is small, it may be possible to allow for any additional flow in the capacity of the pumps when they are first installed. If, however, this development is likely to be extensive, or if its extent cannot be ascertained with any accuracy when the designs are being prepared, it is usually more satisfactory to design the pumping station so that additional pumping units can be added at a later date. This will entail leaving space in both the 'wet well' and the pumping chamber, and may also entail a duplication of the rising main in due course.

When provision is made for any future increases in flow, the effects of the lower flows expected during the earlier years of operation must be considered. If comparatively small flows are held for long periods in the wet well, the long retention time may result in septic conditions developing; this may in turn lead to nuisance from odours at the pumping station, or to problems at the treatment works.

The capacity of the pumping units must be matched by the capacity of the rising main, and by the capacity of any gravity sewers downstream of its point of discharge. If the rising main discharges to a sewerage system, the capacity of this must therefore be checked to ensure that there will be no surcharge when the pumps are operating at maximum capacity. At the *minimum* pumping output, the velocity of flow in the rising main must be sufficient to obtain a self-cleansing velocity; at the same time, if the pumps are of the unchokable type (see below) the pipeline must be large enough to pass any solids which will be passed through the pumps; this usually entails a minimum rising main diameter of 100 mm. It will be appreciated that these considerations limit the *minimum* size of pump which can be installed. The design of rising mains is considered in more detail in Chapter 10.

SURFACE WATER PUMPING STATIONS
As far as possible, pumping stations should be avoided on surface water sewerage systems, in view of the large quantities involved and their comparatively infrequent operation. Mixed-flow pumps are available with outputs up to 1400 to 1500 m^3/h (heads up to 15 m) and are generally considered the most suitable for surface water pumping duties.

Surface water sewage flows are calculated according to the area of the catchment, and a *rate* of rainfall. The figure to be used for rainfall will depend on the area of the catchment (and therefore the 'time of concentration'), and on the economics of the scheme. The *effective* area of the catchment will depend on the type of development or on the type of subsoil, and on the gradients. Formulae and typical calculations for rainfall and surface water flows are given in the author's book on sewerage [25].

SEPTICITY

Sewage turns *septic* when it is stored under anaerobic conditions for any prolonged periods of time. Septicity is accelerated at higher temperatures (above about 20°C). When sewage turns septic, hydrogen sulphide is generated by the action of the anaerobic bacteria on sulphates in the sewage; this can result in the production of sulphuric acid following atmospheric oxidation of hydrogen sulphide which has been absorbed by slime and grease. This acid can be responsible for damage to the sewers and to concrete and metal surfaces of pumping stations, tanks, etc. The generation of hydrogen sulphide also results in offensive odours, while the generation of methane gas (also associated with septicity) can produce dangerous working conditions, especially in sewers and in underground tanks and chambers. Septic sewage is more difficult to treat than is fresh sewage.

Septicity can be prevented or minimised by reducing the time taken for the sewage to reach the treatment works, particularly by reducing delays at pumping stations and in rising mains. Other remedial measures adopted from time to time include aeration of the sewage and the addition of chlorine, lime or nitrates.

The prevention of odours and of septicity in sewage entails good ventilation and good standards of cleanliness. The wet well of a pumping station must be designed to reduce the detention time to the minimum necessary for efficient pump operation. The well must be as self-cleansing as possible; inlet chambers, penstocks and floors of compartments must not permit the accumulation of screenings, grit or scum, and any provision made for the isolation of screenings, etc. must be emptied and cleaned regularly.

The provision of a suitable water point at a pumping station is essential and a supply of hose must be available to reach any part of the building without difficulty. Other essential cleaning equipment includes suitable brooms, mops and squeegees which should be conveniently stored at the site. Adequate washing and toilet facilities for the workmen will encourage the use of cleaning equipment, while a clean and tidy station usually results in 'pride of ownership' by the maintenance staff, with consequent improved standards of maintenance.

LAYOUT OF PUMPS AND OTHER EQUIPMENT

The arrangement of the pumping well (or 'wet well') at a sewage pumping station will depend on the type of pumps being installed.

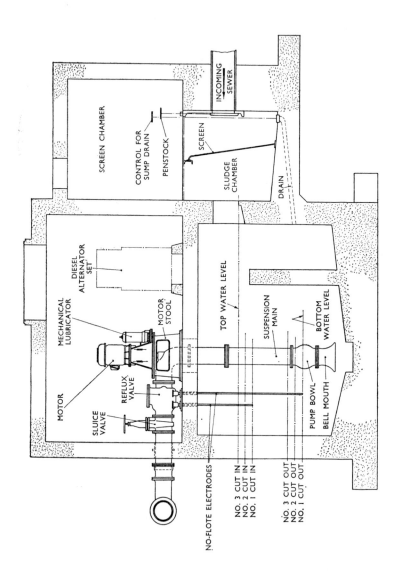

4 SEWAGE PUMPING STATIONS

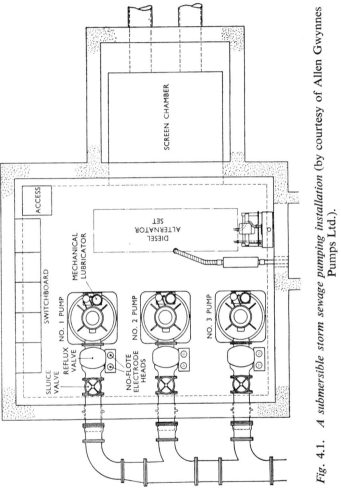

Fig. 4.1. *A submersible storm sewage pumping installation* (by courtesy of Allen Gwynnes Pumps Ltd.).

38　PUMPING STATIONS FOR WATER AND SEWAGE

When submersible pumps are installed (Fig. 4.1), the pumps will be in the wet well itself, while the motors will generally be mounted on a floor vertically above. The more usual arrangement for sewage pumps is illustrated in Fig. 4.2, where vertical-spindle pumps are installed in a dry well alongside the wet well, with the suctions taken through the dividing wall.

While horizontal spindle pumps are cheaper in capital cost, these are not normally practicable, particularly when the sump is deep, *i.e.*

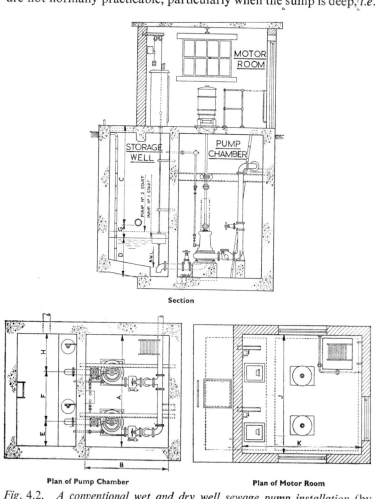

Fig. 4.2.　A conventional wet and dry well sewage pump installation (by courtesy of Adams-Hydraulics Ltd.).

4 SEWAGE PUMPING STATIONS

more than about 3 m. Vertical spindle units occupy less floor space so that the pumping station is cheaper to construct; they are, however, more difficult to install and maintain than horizontal units. A horizontal-spindle installation would usually entail considerably more excavation for both the pumps and the motors, and there is the added possibility of the motors being damaged in the event of flood waters entering the pump chamber.

Other types of installation, including completely submersible pumps (pump *and* motors below water level), ejectors and disintegrator pumps are considered separately in Chapter 5.

Pumping rates can rarely be matched to the rate of flow of the incoming sewage, and provision is therefore made for the storage of the sewage for short periods of time between spells when the pumps are operating. The design of this 'wet well' must minimise the possibility of settlement of solids, and while its capacity must be sufficient to prevent frequent starting and stopping of the pumps, it must be as small as possible to reduce any tendency to septicity. The design of starters is referred to in Chapter 7, while the design of the wet well is considered in Chapter 9.

When pumps are driven by variable speed motors, or by motors designed to run at two or three speeds, it may be possible to reduce the size of the wet well. For these conditions it is usually desirable to allow for the largest individual pump to run continuously for not less than one minute after acceleration to full speed. Many engineers prefer to design on not more than ten starts per hour for each pump even though the switchgear may have a higher rating.

Some engineers prefer to install pumps of varying sizes which are arranged to cut in and out in succession according to the rate of inflow of the sewage, but this is generally considered to be an unsatisfactory arrangement. Not only does it entail the stocking of spares for each size of pump and motor, but as the motors must be electrically interconnected, a fault in the relay system could result in the whole station being out of action. The more satisfactory arrangement consists of a number of pumps of one size (or, at the most, of two different sizes) with two or more pumps operating together when the flow increases. This arrangement has the further advantage that the 'duty' pump can be changed from time to time. The duty pump is the one which starts first, under automatic control, and takes most of the daily load under normal flow conditions. When two sizes of pump are installed, the smaller ones will be used for all normal variations in the 'dry weather' flow, while the larger ones will deal with storm flows.

Where the maximum rate of sewage flow during a day will not exceed about 140 or 150 m^3/h, it is usual to install two pumping units only, each capable of pumping the full flow; each pump then acts as a standby to the other. For larger flows it is normally more satisfactory to install two or more pumps to deal with the normal flows, plus a further pump or pumps as standby. The *minimum* pumping rate must be sufficient to provide a self-cleansing velocity in the rising main of at least 0·6 m/sec.

TABLE 4.1

RECOMMENDED DIMENSIONS FOR A SMALL SEWAGE PUMP INSTALLATION[a]
(*see* Fig. 4.2)

	Pump diameter (*mm*)		
	75/100	125/150	175/200
Dimension A	2700	3600	3900
Dimension B	2250	2850	3450
Dimension C		to suit site	
Dimension D	1050	1200	1350
Dimension E	600	750	900
Dimension F	1200	1500	1650
Dimension G	300	300	300
Dimension H	900	1350	1350
Dimension J Dimension K	depending on details of equipment		

[a] All dimensions in millimetres.

A typical layout of vertical-spindle pumps operating in a dry well is shown in Fig. 4.2; this illustrates a small station with a single duty pump plus one standby. The dimensions shown in Fig. 4.2 can be taken from Table 4.1; these are not the absolutely minimum dimensions and some allowance has been made for walking space and maintenance, but some engineers may prefer to use slightly larger dimensions where appropriate. When the suction pipes are turned down into the sump in a wet well, they should have a clearance of not less than $D/3$, and not more than $D/2$ (where D is the diameter of the suction pipes). This will ensure that the entry velocity is sufficient to scour the floor of the sump.

4 SEWAGE PUMPING STATIONS

When the installation consists of vertical-spindle pumps with the motors at a higher level, the motors are often supported on steel joists which will then form a part of the motor room floor; these joists may also be used to support steel flooring panels. It is preferable to have individual motors sufficiently far apart so that space is available for both the installation and the removal of pumps and pipework. Alternatively, a special access opening may be left for this purpose.

While it may sometimes be convenient to connect the incoming sewer to one end of the wet well (as shown in Fig. 4.2) this is not the ideal arrangement, particularly when dealing with large storm flows, as it tends to induce air-entraining vortices. It will generally be more satisfactory to connect the inlet at the centre of the length of the suction well; this will not only allow for provision of an inlet bay to reduce the approach velocity, but will also allow the division of the wet well into two compartments to facilitate maintenance (*see* Chapter 9).

UNCHOKABLE AND DISINTEGRATING PUMPS

Pumps for use with crude sewage must be capable of passing the solids which can be expected to be found in the sewage. Sewage contains not only sand and grit, but also pieces of rag, wood, metal and, occasionally, dead animals. As the minimum diameter of branch drains in a sewerage system is 100 mm, it is normally assumed that solids of up to 100 mm diameter may be found in crude sewage. Pumps must therefore either be capable of passing solids of that magnitude, or they must be protected by screens or comminutors. Screens should, however, be avoided as far as possible in view of the difficulties in cleaning and in the disposal of the screenings—particularly at isolated pumping stations.

A sewage pump must be easily accessible for quick maintenance and it must be reliable in use; for crude sewage, the pumps must be unchokable. To meet those requirements the efficiency of a sewage pump will usually be appreciably lower than that of a pump designed to handle clean water. Pumps for small and medium capacities must be specially designed if they are to pass 100 mm solids; this will entail the provision of larger clearance between impeller and casing, and, usually, the elimination of any leading eye which can collect rags and other fibrous matter. This has been referred to in Chapter 2. It should be possible to dismantle a pump without removal of sections of pipework; split-casing pumps are usually specified, but handholes may sometimes be considered adequate to allow the clearance of any blockage.

As an alternative to the unchokable type of pump, a disintegrating pump may be used. This type of pump breaks up the solids before they pass through it so that the pump passages can be proportioned to give maximum efficiency. The use of a disintegrating (or macerator) pump enables the use of a smaller diameter rising main; maceration of the solids may also assist in the ultimate treatment or disposal of the sewage, particularly where disposal will eventually be into the sea. This pump can be subject to considerable wear if the sewerage system is 'combined', as the sewage will then contain an appreciable quantity of grit.

The presence of grit in sewage can also cause problems at the glands which seal the annular spaces between the drive shaft and the pump casing. A mechanical type rotary seal may sometimes be used instead of conventional gland packing. A water seal may be obtained either by using clean water at high pressure in the stuffing box, to prevent the ingress of grit, or by incorporating a system of flushing within a mechanical type seal. Grit is present in some quantity in most sewage, and this is the main reason for the lower speeds of sewage pumps as compared with waterworks units; a sewage pump will normally run at a speed of 960 rev/min as compared with 1460 or much more for a clean water pump.

COMPARATIVE COSTS

It is often necessary to compare the cost of a pumping scheme with that of a scheme using gravity sewers; the latter may include a sewer in tunnel or an inverted siphon. Costs to be compared will not only be the estimated capital costs, but also the estimated running costs. While the capital cost of a pumping scheme may be lower than that of an equivalent gravity scheme, it may be found that the running costs will be greater. If so, this would then entail a higher charge on the rates or, if a factory is being sewered, on the running costs of the industry.

The annual cost of any project will therefore be a combination of loan charges (or amortisation costs) on land purchase, civil works and machinery; together with all maintenance costs, including labour and fuel. At present, Government loan charges are based on a loan period of 60 years for land purchase, and for the cost of

4 SEWAGE PUMPING STATIONS

housing, bridges, tunnels and impounding reservoirs; 40 years for other civil engineering works; and 20 years for machinery, plant, fencing and some electrical and mechanical plant. These periods are adjusted from time to time, and, of course, authorities may borrow for shorter periods if they wish. The rate of interest charged on a loan varies from year to year, depending on the financial situation generally and on the current bank lending rate.

The annual cost of a gravity sewer system will therefore be based on any land acquisition charges and the capital cost of the works themselves, using the relevant loan periods and interest rate; plus an estimate of the annual maintenance costs. This figure must then be compared with an estimate of the annual cost of any proposed pumping scheme. The latter will include loan charges on capital works, including land acquisition, pumping station buildings, rising main and any electrical installations, together with estimates of the annual costs of maintenance of machinery (including spares), electrical or other fuel costs, and labour.

There will usually be a number of possible solutions to any problem involving the installation of pumps and rising mains. Rising mains for crude sewage will not be less than 100 mm diameter and the velocity of flow when the smallest pump is operating should not be less than 0·6 m/sec, and preferably nearer 1·0 m/sec; the *maximum* velocity is usually set at about 3·0 m/sec. Within these limits, the solution is then one of economics, as the capital cost of the rising main will increase with its diameter, while the velocity of flow (and therefore the friction head) will decrease. A decrease in the friction head will result in the decrease in the total head; this may affect the size of pumps and motors required, and will certainly affect the power requirements, and therefore the running costs. This subject is considered further in Chapter 10. Usually the most economical *total* head on an unchokable sewage pump is of the order of 18 to 21 m and, where relevant, the rising main should be designed with this in mind.

Probably the most difficult costs to estimate are those for electric power, and those for labour for routine maintenance. Annual power requirements for sewage pumping should be based on the average dry weather flow, plus a percentage for infiltration. If the sewers are on the combined system, allowance must be made for pumping the run-off from the annual rainfall over the catchment area, after a suitable allowance has been made for impermeability (see the author's book on sewerage [25]) and for the effects of any storm sewage overflows. Power charges will be based on a kVA charge according to the installed power of the motors and their power factor, plus a charge for the energy used during the year.

When there is little to choose between two schemes from an economic viewpoint, it should be borne in mind that a gravity scheme will always be simpler to maintain, and will be less liable to breakdown. A pumping scheme should only be chosen if it shows a substantial saving in annual costs.

CHAPTER 5

Some Special Installations for Sewage, etc.

In Chapter 4 the more conventional sewage pumping station was considered. This chapter continues with installations for sewage pumping and looks at the more specialised pumps for small flows and for use at the sewage treatment works.

SLUDGE PUMPING

The layout of a sewage treatment works will usually entail the pumping of the sludge which settles out in humus tanks, and often the pumping of the sludge from the primary settling tanks. In addition, sludge may be pumped from thickening tanks to digestion tanks or dewatering plants. At an activated sludge plant, a proportion of the *activated sludge* will be pumped back to the aeration tanks.

Sludge from primary tanks, *i.e.* 'raw sludge', will contain solids similar to those in crude sewage; the rising main should therefore not be less than 100 mm diameter, and, to obtain a minimum velocity of flow of about 0·75 m/sec, the pump discharge should then not be less than about 6 litres/sec. Humus and activated sludge are less liable to block pipelines and it is generally accepted that a 75 mm minimum diameter rising main will be satisfactory; for a minimum velocity of 0·6 m/sec, the minimum pump output should then be about 3 litres/sec.

One single pumping station may seem desirable to deal with all the 'extra' duties at a sewage treatment works, such as lifting humus sludge to the primary tank or drying beds, mixed sludge to digestion tanks or dewatering plant, and supernatant liquors to the works inlet. It will, however, be apparent that these duties are usually quite different from one another due to the variations in the liquids to be pumped and in the levels of the delivery points. It may also be necessary to give careful consideration to the effect on the efficiency of a small settling tank if a sludge flow of (say) 3 litres/sec is to be discharged to it from time to time.

To allow for the variations in head, many engineers prefer to use plunger pumps for sludge pumping, particularly when the flows are not large. The discharge from these pumps is almost constant over a wide range of heads.

If a centrifugal pump is used for lifting activated sludge, its impeller speed should not exceed about 750 rev/min, as excessive speeds will break up the sludge floc. Provided it is of the correct design and power, a centrifugal pump is usually satisfactory for lifting sludge; it will probably be quieter and more economical than a ram pump. The head is often very low (one or two metres), and an axial flow pump may then be a more satisfactory choice. Screw pumps (*see* Chapter 2), with speeds of 20 to 90 rev/min, are ideal for this duty.

Portable centrifugal pumps with 75 or 100 mm suctions and deliveries, and driven by petrol or diesel engines, are frequently used for lifting sludge and supernatant liquors at sewage treatment works. When these are mounted on hand-barrows with pneumatic tyres they can be moved over the grassed areas and narrow paths which are frequently encountered at small works.

The wet well of a sludge pumping station must be designed as a hopper with steeply sloping sides. If the outlet from the pumps can be located below the level of the bottom of the hopper, there will always be a positive head on the suction side of the pump.

The calculation of the pumping head will be based on the type of sludge and its water content, in addition to the velocity of flow. American practice favours a velocity in the range of 1·5 to 2·5 m/sec; it is generally accepted that for velocities above about 1·2 m/sec, the flow is turbulent and the friction head can be based on standard formulae and tables prepared for the flow of liquids. At lower velocities than 1·2 m/sec, the friction head factor may vary from 1·5 to 4·0 times that used for liquid flow conditions. Formulae and tables for calculating the flow in pipelines are referred to later, in Chapter 10.

SEWAGE TREATMENT WORKS LIQUORS
Within the various sewage treatment processes a number of 'liquors' are produced which must usually be returned to the works inlet for further treatment; this will normally entail pumping. In addition, if the works include biological filters on either the recirculation or

alternating double filtration principles, pumping plant must be installed to handle large volumes of settled sewage.

Supernatant liquors at a works will include those settled out at sludge thickening tanks, wash waters from sand filters and microstrainers, and the liquid contents of storm sewage tanks at the end of a storm. The capacities of pumps for these duties will not be critical within certain limits, as the time taken to return these liquors to a works inlet is almost immaterial. Of more importance is the effect, if any, of these liquors on the treatment process; *maximum* capacities may therefore be set by the treatment capacity of the works units, particularly of the primary settling tanks. A figure of 0·5 d.w.f. is often quoted as being the maximum advisable increase in the flow due to the pumping back of supernatant liquors.

While in theory centrifugal pumps suitable for clear water duties could be used for these liquors, in fact there is always the possibility of sludge or floating matter being pumped along with the liquid; the pumps should therefore preferably be of the unchokable type.

It may be convenient to combine the supernatant liquor pump duties with those of the sludge pumps at a works. This has been referred to above, and the relative duties must be examined carefully to ensure that any combination of duties is satisfactory.

When pumps are installed for the recirculation of effluent or for an alternating double filtration scheme, the capacities will depend on the design of the particular scheme. For a recirculation scheme the *minimum* quantity of effluent returned will often be 1·0 d.w.f.; the maximum may be up to 2·0 or 2·5 d.w.f. For operation on a.d.f., the pumping rate will vary with the flow of raw sewage to the works, and the rate may be between 0·5 and 3·0 d.w.f. For either system, the number and sizes of the pumps must provide a sufficient number of steps between minimum and maximum flow rates; control will normally either be by floats or electrodes, and should be regulated by the rate of flow of the incoming raw sewage. As the flow of sewage increases in a recirculation scheme, so the amount of effluent recirculated will be decreased. Four equal capacity pumps can usually cover a range of duties from 1·0 d.w.f. to about 2·5 d.w.f., depending on the friction head in the rising main.

SUBMERSIBLE PUMPS

When pumps are installed for pumping storm sewage *only,* these may sometimes be more conveniently of the vertical-spindle submerged

48   PUMPING STATIONS FOR WATER AND SEWAGE

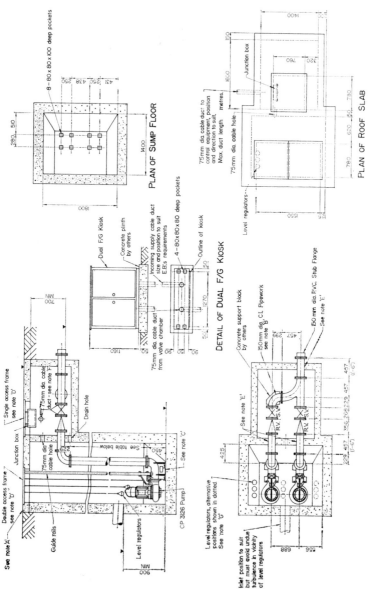

Fig. 5.1. A submersible sewage pumping installation (by courtesy of ITT Flygt Pumps Ltd.).

5 SOME SPECIAL INSTALLATIONS FOR SEWAGE, ETC.

type, with the motors installed at a higher level. These pumps may be of the mixed-flow type as they are not required to pass larger solids. Use of this type of pump (as compared with a 'traditional' vertical-spindle installation with wet and dry wells) reduces the size of the pumping station by the omission of the dry well; at the same time the installation of the pumps below the liquid level ensures that they are always fully primed for use. A typical installation is illustrated in Fig. 4.1.

As storm sewage pumps are not designed to run for long periods, their maintenance can be less frequent than for a crude sewage pump. When maintenance of a submersible pump is required, however, the complete unit must be withdrawn from the wet well.

Units with close-coupled pumps and motors, and designed to run completely submersed, have been available for many years in the water supply field (*see* Chapter 3), and also for temporary pumping installations. More recent developments have produced completely submersible sewage pumping units for permanent installations; these are usually fitted with quick-acting couplings to facilitate the removal of the pump for maintenance (*see* Fig. 5.1). Pumps are available with discharge rates up to about $1700 \text{ m}^3/\text{h}$ and for heads up to 35 m; they are normally of the unchokable type, while disintegrating pumps are available for the smaller duties.

SUMP DRAINAGE

Provision is usually made for pumping out small quantities of water from the dry well of a pumping station. There should not be any leakage into the dry well either from the wet well or from the surrounding ground if the concrete work is properly constructed, but there will be slight leakage from the glands of the pumps themselves, and some condensation water can usually be expected in underground chambers. The floor of the dry well should be graded to a small sump to collect this water; this can then either be fitted with a small automatic sump pump operated by a float or electrodes, or auxiliary suctions can be fitted to two or more of the main pumps.

An independent sump pump is preferred by many designers particularly if the main suction pipes are 200 mm or more in diameter. This should be a pump suitable for dirty water (but not sewage), and at a sewage pumping station the small diameter rising main can be arranged to discharge into the wet well. Small self-contained submersible units are available for this type of duty (*see* Fig. 5.2). When a cellar drainage pump is used the motor should be mounted at least 500 mm or so above the floor.

When auxiliary suctions are used, they should be connected to the horizontal section of the main pump suction between the pump and

Fig. 5.2. *A submersible sump pump* (by courtesy of James Beresford and Son Ltd.).

the suction pipe sluice valve. The auxiliary suction pipe should be at least 40 mm diameter, should be as direct as possible, and should be provided with a 'fullway' type valve with its spindle extended for operation at the motor-room floor. At least two of the pumps in any pumping station should be provided with auxiliary suctions if an

5 SOME SPECIAL INSTALLATIONS FOR SEWAGE, ETC. 51

independent sump pump is not installed. Care must be taken to *close* the valves on the auxiliary suctions after use, to avoid flooding the dry well.

PNEUMATIC EJECTORS

A pneumatic ejector is a convenient tool for lifting small quantities of sewage without the need for screens or the provision of a suction well (*see also* Chapter 2). The sewage being completely enclosed, there is little or no possibility of smell nuisance. Ejectors are reliable in use and their maintenance costs are low; their efficiencies are, however, very low as compared with centrifugal pumps, particularly at the higher heads, and they are generally more expensive than pumps in capital cost.

The usual installation will consist of an automatic air compressor, together with air storage 'bottle', and the ejectors themselves. At least two ejectors should be provided so that a standby is available during repairs and maintenance; a second compressor should also be provided if a breakdown would have serious results.

An ejector is rated according to the capacity of the cylinder, *e.g.* 140 litres, 250 litres, etc. As the discharge is normally assumed to be completed in one minute, these ratings would then be 140 litres/min, 250 litres/min etc. The *actual* discharge rate will be well in excess of its rated discharge (normally about double), particularly if the air supply is on the generous side. It will be seen then, that for a 100 mm diameter rising main it would be possible to use a nominal 8·4 m^3/h (140 litres/min) ejector as the actual output would be nearer 17 m^3/h, and the velocity of flow in the main would then be almost 0·6 m/sec.

The standard type of ejector is installed *below* the level of the sewer so that it fills by gravity. Where excavation would be unduly expensive it may be preferable to install 'lift and force' type ejectors (Fig. 2.7). These are more convenient for inspection and maintenance and there is less risk of foreign bodies becoming wedged in the inlet valve.

PACKAGED PUMPING UNITS

The packaged pumping station is a recent development for the smaller type of installation. These are available from a number of manufacturers and with outputs of up to about 200 m^3/h. Some types are supplied in factory-built chambers ready for installation with very little civil engineering work; a separate wet well may be provided. Other types are based on a welded steel cylindrical 'wet well' to which are attached pumps, starters, pipework etc.; this complete unit is then installed underground in a simple brick or concrete chamber (*see* Fig. 5.3).

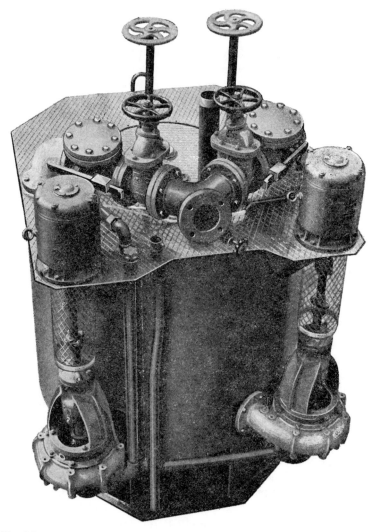

Fig. 5.3. A packaged sewage pumping unit (by courtesy of Pegson Ltd.).

5 SOME SPECIAL INSTALLATIONS FOR SEWAGE, ETC.

When the pumping set incorporates unchokable pumps and is installed for use with crude sewage, the rising main must be at least 100 mm diameter. The *minimum* output will then be controlled by the velocity of flow in the rising main. As discussed in Chapter 4, the velocity of flow for crude sewage should not be less than 0·6 m/sec and should preferably be nearer 1·0 m/sec. For a 100 mm pipeline, these figures represent discharges of about 4·85 litres/sec (17·5 m^3/h) and 8·11 litres/sec (29 m^3/h) respectively.

Totally submersible pumps and motors are ideally suited for use with a packaged station in an underground chamber. A superstructure may be preferred for the switchgear, but in some cases this can also be installed below ground.

SMALL FLOWS

Some of the problems encountered with small flows of sewage have already been considered in relation to the installation of ejectors or packaged units. Probably the most important of these is the need to maintain a self-cleansing velocity in the rising main and yet, at the same time, to pump the sewage sufficiently frequently to avoid any problems of septicity.

Where the peak flow in the sewerage system will not normally exceed six times dry weather flow, and where this will not be greater than about 130 or 140 m^3/h, a pumping station will normally consist of one pump for use during all variations in flow, with a second similar pump as standby. For larger installations three or more units will be installed. Where the sewage is neither screened nor macerated, the usual minimum pump capacity must be about 20 m^3/h to give a self-cleansing velocity in a 100 mm diameter main; it will be found, however, that few manufacturers are prepared to market a pump as 'unchokable' and suitable for solids up to 100 mm, with an output of less than about 50 or 60 m^3/h.

This limitation has been overcome to some extent in the Pulsometer 'Solids Diverter'. In this equipment, the solids are diverted around the pump which then lifts comparatively clear water only. With this arrangement the pump can be of higher efficiency and the limitation on the equipment is often the velocity of flow in the rising main. Solids Diverters are manufactured for maximum incoming sewage flows of from 6 m^3/h to 34 m^3/h; the discharge rates of the diverter vary from 27 to 70 m^3/h respectively.

When a pump operates in conjunction with a built-in disintegrator, the diameter of the rising main can be reduced below the previously recommended minimum of 100 mm; pumps with outputs as low as 2 or 3 m^3/h can then be used along with a rising main of 30 or 40 mm diameter. The small diameter of the main reduces the cost of a scheme and also the overall capacity of the main itself; this reduces the retention time of the sewage in the main and therefore the possibility of septicity. With the Mono 'Mutrator' (*see* Fig. 5.4), the sewage is vacuum lifted so that any unacceptable solids such as stones or large metal objects are rejected back to the sump. A macerating device in the suction line of the pump then reduces the size of all particles of paper, textiles, rubber, etc. before the sewage is passed through the

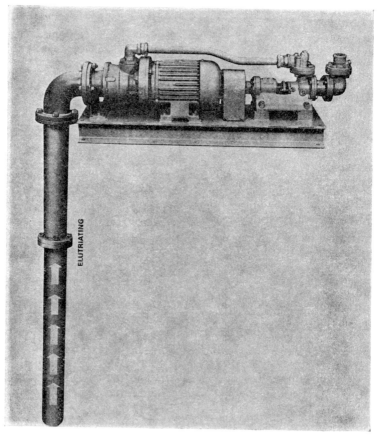

Fig. 5.4. A disintegrating pump (by courtesy of **Mono Pumps Ltd.**).

5 SOME SPECIAL INSTALLATIONS FOR SEWAGE, ETC.

pump. This pump is available with an output as low as $2 \text{ m}^3/\text{h}$, for use with a 30 mm diameter delivery pipe.

METERING PUMPS

Metering pumps can be installed when accurate measurement is required. They are usually either of the diaphragm type or the plunger type (for higher heads). They are more expensive than normal pumping units but they are useful in certain circumstances—particularly in pilot scale plants and laboratories, and in treatment involving dosage with chemicals.

Arrangements can be made for the output of a metering pump to be varied manually at the pump, or by remote control, and either at rest or in motion. Automatic variation of output is possible.

PORTABLE PUMPS AND TEMPORARY INSTALLATIONS

Portable pumps as used by contractors are well known; these may now have either petrol or diesel engine drives. They have been developed from the successful steam pumps of the nineteenth century. The traditional diaphragm and ram pumps have stood the test of time and are still very popular on smaller contracts. More recent types of pump include centrifugal and totally submersible pumps; the latter, of course, can only be used when a supply of electricity is available. The modern contractor's centrifugal pump has been developed since the advent of well-point dewatering, and the portable pump with vacuum priming has become almost universal in the larger sizes; this type of equipment will run for many hours without attention. At present most larger pumps on contractors' sites are either vacuum assisted centrifugal pumps or totally submersible pumps.

At a sewage treatment works the portable pump has many uses; these include the pumping of sludge and the emptying of tanks. They are also used on water and sewerage schemes in temporary installations during extensions, repairs, etc. Emergencies such as flooding after a storm, or a fractured pipeline, may entail pumping large quantities of liquid for a comparatively short period. Where crude sewage or other liquid containing solids is to be pumped, the pumps should be of the unchokable type and the suction and discharge pipes should be at least 100 mm diameter.

CHAPTER 6

Prime Movers

Most modern water or sewage pumps in permanent installations are driven by either direct-coupled electric motors or by internal combustion engines. Electric motors will generally be more economical in capital and running costs unless power is not already available at the site, or if pumping will only be required at infrequent intervals (*e.g.* for pumping storm sewage). Very limited use is made of wind and water power for smaller installations. The use of steam has been superseded, and is now only of historical and academic interest.

In addition to the relative economics, other considerations may include the effects of power failure or fuel shortage, reliability, efficiency and quietness in operation. It may sometimes be convenient to provide electric motors for normal duties, with an internal combustion engine as a standby. Internal combustion engines can be arranged for direct drive of the pumps or they may be coupled to generators, which in turn, will drive electric motors.

ELECTRIC MOTORS

The electric motor is a convenient and reliable prime mover for all forms of pumping. It is particularly suitable for operation in an unattended station, as modern automatic controls are very reliable.

With an ever-widening high voltage grid network and increases in the number of generating stations throughout the country, the possibilities of power failure are now much less than a few years ago. Two sources of supply may sometimes be justified as a precaution against power failure, while for large stations, a high voltage supply (6600 V or more) may result in a better tariff and in a reduction in restrictions on starting loads.

The choice of electric prime movers will generally be between slip-ring and squirrel-cage motors. The latter are cheaper in capital cost

6 PRIME MOVERS

and are usual for small and medium installations, particularly if high starting currents are acceptable. Small squirrel-cage motors are often controlled by direct-on-line starters; larger installations normally have star-delta or auto-transformer starters (*see* Chapter 7).

Slip-ring motors, with stator-rotor starters, are frequently used for both water supply and sewerage installations; they give full-load torque on starting (for efficient operation of centrifugal pumps) with about 1·25 times full load current, and they are very suitable for automatic control. When employed with rotor-resistance control, a slip-ring motor can be used as a variable speed motor so that the pump output can be varied. Variation in speed of the motor below its design (synchronous) speed must, of course, result in a reduction in its efficiency; this will, however, usually be less than any loss of efficiency which would result from the throttling of a valve to reduce a pump's output by a similar amount.

Squirrel-cage motors (short-circuited rotor motors) are simple in design and their efficiency will generally be equal to corresponding slip-ring motors. Direct-on-line starters are the most satisfactory to provide the torque required to start centrifugal pumps, but, as the starting current is then about six or seven times normal full load current, for larger motors this is normally not economical nor may it be acceptable to the electricity authority. Squirrel-cage motors are therefore usually installed with star-delta or auto-transformer starters; the starting torque may, however, then be down to between 33% and 80% of the full-load torque. 'High torque' motors are now available for use where slip-ring machines would previously have been necessary; these have starting torques of about double those of standard squirrel-cage motors.

Variable-speed a.c. commutator motors with induction regulators are less efficient than either of the above type of motor, but are useful where variation in pump output is an important aspect of an installation. They are economical in larger installations. Various types of these motors were described by Summerton in 1961 [18].

The synchronous motor with a power factor of unity has a high efficiency, and is suitable for large installations using alternating current. These motors have fixed speeds and cannot be used for variable speed duties. While more expensive in capital cost than squirrel-cage motors, synchronous motors are normally cheaper than slip-ring motors of similar size.

When d.c. current is available, the d.c. motor has some advantages,

It has good starting characteristics and high efficiency, and can be used for variable-speed duties while maintaining its efficiency over a wide range. The use of d.c. current is not now very common, and it is usually limited to use with diesel-driven generators.

The submersible motors used either in conjunction with submersible borehole pumps and some types of booster pumps, or with submersible sewage pumps, are normally three-phase squirrel-cage induction motors using either direct-on-line or auto-transformer starters; single-phase motors may be used for the smaller sizes. In view of the large ratio of length of rotor to the diameter, borehole motors are generally less efficient than conventional motors.

In his paper in 1961 [18], Summerton stressed the need for the designer of a scheme to limit the speed of operation of electric motors. He pointed out that 'the higher the speed, the greater the risk that the plant will be noisy, due chiefly to the windage of the fan on the electric motor', and he suggested that 'borehole pumps with vertical shafts should be limited to 1000 rev/min, and horizontal, centrifugal and turbine pumps to 1500 rev/min, with a 15% margin upwards if variable-speed motors are used'.

The speed of rotation of a d.c. motor is more-or-less infinitely variable, and is only limited by mechanical considerations. With a.c. motors, the choice of speed is set by the current frequency, and by the number of pairs of poles in the motor, *e.g.* for a 50 Hz current and two pairs of poles (four poles), the nominal speed of rotation will be:

$$\frac{50 \times 60}{2} = 1500 \text{ rev/min}$$

In fact, under full load, slip will usually reduce this figure to about 1440 or 1450 in smaller motors.

For both water supply and sewage pumping it is usual to use drip-proof motors in indoor situations; totally enclosed motors are sometimes used, and these *must* be used when installed out of doors. Submersible motors are, of course, specially designed to work under water; they make use of the water to assist in cooling, and in the wet stator type the water circulates through the windings (suitably insulated).

Both squirrel-cage and slip-ring motors are induction motors, and as inductance causes a lagging power factor, this is a feature of most electric motors. Power factor is the ratio of the useful current in a circuit to the total current being used, and it therefore represents a loss, as cable sizes must be large enough to reduce heat losses, and substations and other apparatus must be rated for the larger current. The power factor of both squirrel-cage and slip-ring motors is usually in the range of 0·75 to 0·90 and may be as low as 0·65. When the power factor is less than about 0·90, a penalty charge may be imposed by the electricity authority. Power factor reduction can be avoided by running motors at or near their rated duties, while power factor correction can be obtained by the installation of a capacitor between the starter and the motor (see Chapter 7).

The Area Electricity Boards provide an Industrial Advisory Service which will assist in the choice of tariffs, and will provide advice on load and power factor control.

INTERNAL COMBUSTION ENGINES

As an alternative to electric motors, internal combustion engines are very adaptable, as variation in speed is generally possible. Internal combustion engines use either petrol, oil, or gas; engines using gas as a fuel may operate on gas alone, or on a mixture of oil and gas.

It is generally accepted that where electric power is continuously available, electric motors will be more economical in both capital and running costs. The internal combustion engine becomes a viable proposition when power is not available or when pumping is only required at infrequent intervals, e.g. for flood control works and for pumping storm sewage and surface water. In comparing costs of proposed installations, the initial estimated capital cost must include all installation costs, together with the costs of any special foundations, while the estimated running costs must take into account maximum demand and power factor charges for electric power, or handling costs and the skilled labour for other fuels.

Petrol engines have limited application for water and sewage pumping in view of the high fuel and maintenance costs, but they may be used as standby units, or for very small installations, especially at sewage treatment works. They are generally expensive to run and they require fairly close supervision. The more usual internal combustion engine in use today uses heavy oil or diesel, or some form of gas. Oil engines are popular for waterworks use and for some types of sewage pumping (particularly as standby units for use in the event of power failure). Straight gas engines are used to a very limited extent; at some larger sewage treatment works where gas is available from the sludge digestion process, dual-fuel or alternative fuel engines may be economical. Engines may be coupled to the pumps, either directly or through gearing, or they may be used to drive generators.

Internal combustion engines can be arranged for manual or automatic start, and also for manual or automatic stopping. Automatic operation is generally only employed when relatively low power is involved. Where units are installed as standby sets in fully automatic pumping stations, the internal combustion engines will themselves normally be arranged for automatic control. Where relevant, automatic clutches can be installed so that a pump normally driven by an electric motor can be driven by an internal combustion engine in the event of power failure. Provision must be incorporated to stop the starter motor if a diesel engine fails to start in (say) half a minute; further safety devices are usually incorporated to stop the engine in the event of faulty lubrication or overheating.

A diesel engine for water or sewage pumping duties is normally air-cooled and complete with a flywheel. Equipment to be installed with the engine will generally include a service fuel tank, a lubricating oil filter, starting equipment (including an independently-driven compressor or battery), and provision for charging the battery directly from the diesel engine. The design of the station must allow the engine to operate each time for a sufficiently long period (not less than 15 min) to allow it to recharge the starter battery. An alternative mains-operated trickle charger should also be installed. As fuel oil must be delivered in bulk by tanker, provision must be made at the site for fuel storage.

Where internal combustion engines are used for normal duty pumping (as opposed to standby duties), it may be possible to salvage some of the waste heat for space heating or (at a sewage works) for sludge heating. If so, allowance should be made for this heat recovery when comparing estimated running costs of alternative proposals.

STEAM POWER

While the use of steam power was quite common up to the years immediately following the 1939–45 War, its use has decreased considerably in recent years. Steam engines are still used but they have very little significance in water or sewage pumping. Steam turbines may be competitive for large waterworks undertakings; these will usually be of the impulse type.

COMPRESSED AIR

Compressed air is used to operate air lift pumps used in both water supply and sewerage, and also in the compressed air ejectors used for smaller sewage flows.

When used in a well or borehole the air lift pump has a number of advantages; it has no moving parts and therefore its efficiency is not impaired by grit or mud, the air helps to purify and cool the water and one compressor can be used to operate the pumps at more than one well. The overall efficiency of an air pump is only normally between 25 and 50%, and its output is limited by the depth of immersion of the air pipe (as compared with the operating head), and by the relative cross-sectional areas of air pipe and rising main. The principal advantages of air lift pumping are simplicity and ease of maintenance.

The accepted formula for volume of air required for air lift pumping is:

$$V_a = \frac{h}{C \log [(H + 10 \cdot 4)/10 \cdot 4]}$$ Formula 6.1

where

V_a is the volume of free air, in m^3/m^3,
h is the total lift, in metres,
H is submergence when operating, in metres,
C is a constant.

Values of C and recommended submergence of the air pipe below water level in the well are given in Table 6.1. Air must be supplied at a pressure of about 10 kN/m^2 per metre depth of submergence.

TABLE 6.1
VALUES OF C
AND
RECOMMENDED SUBMERGENCE OF AIR PIPE FOR AIR LIFT PUMPING

Lift (m)	C	Submergence required (ratio of submergence to lift)
15	10·2	2·5
20	9·7	2·2
25	9·7	2·0
30	9·6	1·8
40	9·4	1·6
50	9·2	1·4
60	9·0	1·3
75	9·0	1·2
100	9·0	1·1

Compressed air is used to operate sewage ejectors. These are very suitable for lifting the sewage from small communities or from isolated buildings, and have been described in Chapters 2 and 5.

WATER POWER

Water power may be employed in some circumstances in the form of turbines coupled to pumps or generators, when surplus water head is available. For smaller installations, the hydraulic ram is a useful machine where the conditions are appropriate (*see* Chapter 2).

WIND POWER

Wind power is used to drive pumps in some rural installations but there is a tendency for this to be replaced by small petrol or paraffin engines.

SHAFTING AND COUPLINGS

In vertical-spindle installations, the electric motor is normally mounted vertically above the pump. The two units may be directly-coupled or extension shafting may be used to allow the motor to be installed at ground level. The connection between motor and pump or between shafting and pump should be by means of muff or similar couplings so designed as to permit their disconnection without disturbing either the motor or the pump.

Shafting may sometimes be possible without any intermediate bearings. These bearings may however be necessary to prevent 'whip' in the shafting; if so, supporting joists must be provided. Any exposed shafting should be protected by sleeves or wire cages.

The use of Hardy Spicer couplings for connecting sections of extension shafting permits the coupling up of motors and pumps which may be slightly out of line; use of these couplings avoids the necessity of very accurate adjustment during installation.

CHAPTER 7

Pump Starters, Controls and Other Accessories

This chapter looks at the various forms of starters and other controls used in association with pumps driven by electric motors. Particular reference is made to the automatic control of both sewage and water supply pumps. A series of papers on the automatic operation of waterworks plants was published in 1956 [13]; the use of 'telemetry' was considered at a Symposium in 1967 [20].

STARTERS

Direct-on-line starting may only be permitted by the electricity supply authority for comparatively small motors when these are used in isolation. When an installation includes a number of motors, the starting current then becomes a relatively small part of the total connected load and is of less importance. When power is generated privately (in the mining industry, for instance), a limit may not necessarily apply; direct-on-line starters are available for use with comparatively large motors. The starting current with a direct-on-line starter may be from five to seven times the full load current at a low power factor, and this may be important when the electric power tariff includes a kVA charge based on the magnitude of the starting current.

The more usual type of starter for use with centrifugal pumps will be either of the star-delta or of the stator-rotor type. The starting current of a star-delta starter (used with a squirrel-cage motor) will be about 100 to 250% of the normal full load current. With stator-rotor starting and slip-ring motors, the starting current will be about 125% of the full load current.

As discussed in Chapter 6, a slip-ring motor with a stator-rotor starter gives a starting torque equal to full load torque, and in that respect, it is comparable with direct-on-line starting. A starting torque of between 60 and 100% full load torque is desirable with pumps which may stand idle for some time, to overcome any sticking or minor blockage which may occur. This is particularly important with sewage and land drainage pumps.

The ratio of initial to full voltage in normal star-delta starting cannot be varied; the starting torque is proportional to the square of the voltage. Auto-transformer starting for squirrel-cage motors is more expensive but is preferable as, with a Korndorffer connection, this provides for some variation in the starting voltage (usually 50%, 65%, 75% and 85% of the full voltage).

Starters are rated according to their proposed duty. 'Ordinary Duty' rheostat starters are rated at two starts per hour, 'intermittent duty' starters are rated for up to 15 starts per hour, while 'heavy duty' equipment is suitable for 40 starts per hour. In sewage pumping station design it is normal to design for a maximum of 15 starts per hour, although engineers sometimes specify heavy duty starters to provide an added factor of safety to allow for development. Alternatively, the design may be based on seven or eight starts per hour, using intermittent duty (15 starts) equipment. For intermittent duty starting, the storage capacity in a pumping well between 'start' and 'stop' levels (*see* Chapter 9) must be equal to the output of the largest pump *per minute,* or:

$$Q = \frac{D}{60}$$ **Formula 7.1**

where

Q is the storage capacity, in m^3,
D is the pump discharge, in m^3/h.

Formula 7.1 is also used to determine the space between starting levels of individual pumps or of the different speeds of multi-speed pumps.

SWITCHGEAR AND ACCESSORIES
Starter cubicles may be of the floor-mounting type or they may be for wall-mounting. Larger installations are usually built up into a composite floor-mounting panel to include cubicles for switchgear and recorders, together with units to house fuses and circuit breakers of subsidiary circuits. A typical composite panel is shown in Fig. 7.1.

7 PUMP STARTERS, CONTROLS AND OTHER ACCESSORIES

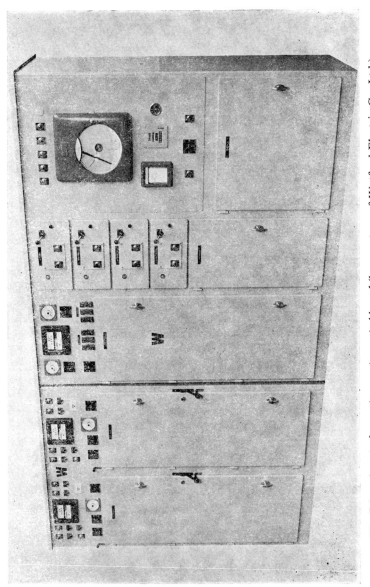

Fig. 7.1. A typical pumping station switchboard (by courtesy of Watford Electric Co. Ltd.).

Wall-mounting units are generally cheaper and are suitable for smaller pumping stations housing only one or two pumps. Units for external use can be either floor or wall-mounting; in either case they must be fully weatherproof. Each starter cubicle should incorporate a suitable main isolating switch, operated from outside the cubicle, and mechanically interlocked with the door of the cubicle so that the door cannot be opened when the current is 'on'.

Push-button control can be provided—the push-buttons providing an automated 'hand' start and stop. Additional remote push-buttons can be incorporated, but for safety reasons, there must always be one 'master' stop button; this is usually mounted at the starter or at the motor itself. When pump motors, or other motors, are provided with automatic start and stop devices, it is usual to include a 'hand-off-auto' switch to provide local control for hand operation.

Incorporated into the starter units will be 'no-voltage' and overload releases, isolating switches and change-over switches, fuses and circuit breakers. An ammeter is usually included, together with red and green 'running' and 'off' lights. Other items in a cubicle may include a voltmeter, starter relay units for automatic control, kWh meters, alarm lights and bell, battery and controls for any low voltage emergency lighting, and a clock. Any hydraulic meters, pumping station heating and lighting circuit controls, and the controls for any external circuits may also be included where convenient. Other 'optional extras' include an 'hours run' meter for each pump, time lag relays where appropriate and thermostatically-controlled panel heaters.

METERS, RECORDERS AND TRANSMITTERS

Meters and recorders have many uses in water supply and sewerage schemes, and it is usually convenient to provide transmitters so that the information can be repeated at a distance from the event; a pumping station often forms a suitable central point for these instruments. The rationalisation of water sources, 'together with the rapid growth of electronic telecommunication, is transforming the nature of the control of pumping machinery' (I.W.E. Manual [32]).

On water schemes, 'telemetry' can form a vital part of the system, enabling information on flows and levels to be transmitted to a control point and providing alarms for low and high water levels in reservoirs, in addition to the remote control of pumping equipment. Digital or analogue computers can be used, in conjunction with Post

7 PUMP STARTERS, CONTROLS AND OTHER ACCESSORIES 67

Office telephone lines or radio links [20]. River authorities use telemetry for data collection, and for monitoring and warning systems, as well as for the automatic operation of river regulation systems. The transmission of information on a number of measurements can be carried out over one telegraph or radio link simultaneously by frequency-division multiplexing, or automatically in turn by term-division multiplexing.

Hood [7] reports that in Sydney each new sewage pumping station is 'being equipped with a signal alarm system which connects to a central control by electronic-line telemetry equipment. With this system a breakdown is immediately relayed to the control, and the nature of the breakdown is indicated'.

When information is to be transmitted automatically by telemetry, the measurands (the measured variables) must first be converted into a suitable electrical form in a 'transducer'. The types of transducer available for the various parameters involved include equipment for use with orifice plates, Dall tubes and Venturi tubes. In addition, equipment is available for use with magnetic flow meters and for the transmission of information on pressure variations, temperatures, pH and conductivity.

On sewerage systems and at sewage treatment works the operation is to some extent more localised. Information on the level of the liquid in a channel or tank, or the magnitude of flow as recorded by a flow meter, can be used to open or close valves and penstocks, or to control pumps. When effluent is recirculated, the operation of the recirculation pumps can either be controlled by liquid levels or from a rate-of-flow meter.

CONTROL BY LEVEL AND PRESSURE TRANSMITTERS

Control by some form of float gear is used extensively in the waterworks industry. Due to possible fouling of cables and pulleys at sewage pumping stations, float gear is being superseded in sewerage practice by other forms of control. Level controls can be used to stop and start pumps when the liquid level reaches predetermined maximum and minimum levels in a receiving reservoir, or according to the levels in the pump suction well. The relays are usually connected to a plug and socket board or other switching arrangement so that any pump can be chosen as the 'duty' pump. A sewage pumping station will normally be arranged so that the first pump to start will be the last to cut out. When necessary, starters can be fitted with suitable adjustable time delay devices so that there is no possibility of more than one motor starting in a station at one time.

The simplest float-operated level transmitter consists of a float suspended by a cable which runs over a pulley. Variations in the level of the float are transmitted through the pulley and appropriate gearing to operate a potentiometer or other transmitting element. Alternatively, the floats may slide up and down on rods which themselves are moved by the pressure of the floats on adjustable stops. Various modifications can be incorporated to increase the sensitivity and accuracy when this is necessary.

Sewage pump control is now often arranged with electrodes; these are probe-type detectors of metal which use the liquid as an electrolyte to complete an electrical circuit to operate a magnetic relay (*see* Fig. 7.2). Special 'low sensitivity' controllers are available to combat

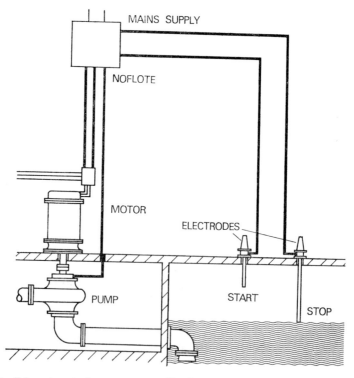

Fig. 7.2. *A typical arrangement of level electrodes* (by courtesy of Fielden Electronics Ltd.).

7 PUMP STARTERS, CONTROLS AND OTHER ACCESSORIES

problems caused by the fouling of electrodes by rags, etc. Electrodes are usually arranged in pairs as 'stop' and 'start' electrodes for each pump, with, if required, additional electrodes to operate alarm

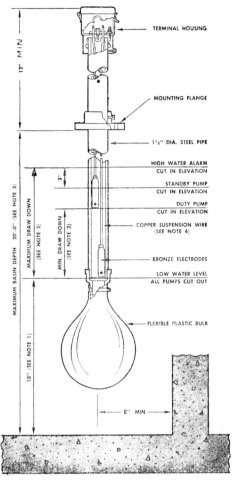

Fig. 7.3. Sealed floatless pump control (by courtesy of Varley–FMC Ltd.).

circuits, etc. One patented form of control uses only one pair of electrodes for a number of pumps, in combination with motor-driven controllers to operate 'increase pumping' and 'decrease pumping' timers and relays [1]. In the 'Sealtrode' system illustrated in Fig. 7.3, the electrodes are completely enclosed; the level of the electrolyte within the tube is varied according to the hydrostatic pressure on the flexible bulb.

Mercury switches enclosed in casings of polypropylene can be used to control the operation of submersible units for sewage pumping, drainage, etc. These are simple pear-shaped units suspended from their own cables; as the liquid level rises, the casing tilts to allow the

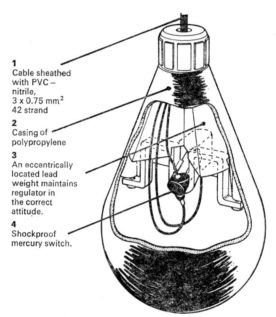

1 Cable sheathed with PVC – nitrile, 3 x 0.75 mm² 42 strand

2 Casing of polypropylene

3 An eccentrically located lead weight maintains regulator in the correct attitude.

4 Shockproof mercury switch.

Fig. 7.4. *An enclosed mercury switch* (by courtesy of ITT Flygt Pumps Ltd.).

mercury switch to close a circuit either to start or stop a pump, or to operate an alarm. Units suitable for liquids with a specific gravity range from 0·65 to 1·50 are illustrated in Fig. 7.4.

A pneumatic bubbler, operated by compressed air, can be used to provide an indication of the water level in a tank or channel. When a chart-type level recorder is installed, pump 'stop' and 'start' can be arranged by the addition of micro-switches tripped by the stylus arm of the indicator. This is shown diagrammatically in Fig. 7.5.

Electrodes may be used in boreholes to provide 'no water' alarms and cut-outs; a transducer working with a differential transformer

7 PUMP STARTERS, CONTROLS AND OTHER ACCESSORIES

transmitter can also be used to operate alarm signals, or for level recording and the automatic operation of pumps.

Whether electrodes or floats are used in a pump suction well, they must be sited carefully to avoid the effects of vortex action when a pump is operating. Starting levels between individual pumps should be at not less than 150 mm intervals, to allow the first motor to

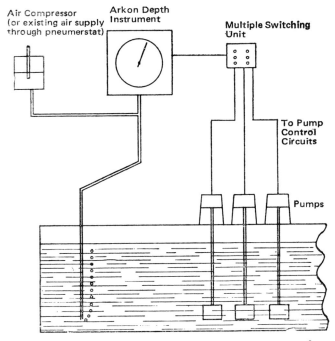

Fig. 7.5. Using a multiple switching unit (by courtesy of Arkon Instruments Ltd.).

develop its full speed and output before the second motor starts. This is particularly important when using electrodes, where wave action may cause 'false' starts and stops. Electrodes or floats can also be used for the operation of variable speed or multi-speed motors or to vary the pitch of pump impellers.

Booster pumps in a water supply system can be arranged to start when the level in a reservoir or elevated tank reaches a predetermined level, and to stop at another level. Where no reservoir is provided, the pumps can be started and stopped by means of pressure-operated

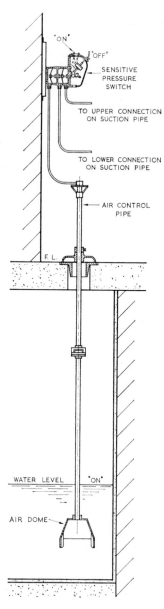

Fig. 7.6. The 'Brownson' floatless control switch (by courtesy of Tuke and Bell Ltd.). *This switch is specifically designed for use with Tuke and Bell equipment.*

7 PUMP STARTERS, CONTROLS AND OTHER ACCESSORIES

switches of the diaphragm type. In some cases, a combination of reservoir level 'start' and pressure switch 'stop' may be convenient.

The pneumatic switch can be either of the sealed type, mounted outside the tank or source of supply, or an open screen type for immersion in the liquid. Each type consists of a top and bottom casing, separated by a flexible moulded diaphragm. The lower part of the casing is connected to the liquid, while the upper part is connected to the measuring instrument. The flexible diaphragm is displaced by the head of the liquid until the air pressure above the diaphragm corresponds to the liquid pressure. Periodic flushing of the connecting pipe through the wet well wall is necessary to ensure trouble-free operation. Pneumatic type controls are used extensively in the United States, but are not so common in the United Kingdom. The patented 'Brownson' floatless control switch (Fig. 7.6) is operated pneumatically by means of an open-ended air bell suspended at the end of an air control pipe; this pipe is connected to a switch unit incorporating primary and secondary diaphragms, connected to a lever carrying a mercury switch.

CUT-OUTS, ALARMS AND INDICATING LIGHTS

Cut-outs are fitted into starters to protect the equipment against overload in case of a fault developing. High-rupturing-current (H.R.C.) fuses are available which allow momentary surges (such as occur during the starting of a motor) but which operate within a quarter of a cycle. It may sometimes be economical to install an isolating switch with H.R.C. protecting fuses in lieu of a main circuit breaker, but once fuses have operated they must be discarded and replaced. The normal provision against overload in a starter cubicle is some form of trip which will operate slowly under small overloads but which will operate in 1·5 to 2·0 cycles under heavy fault conditions. These trips can be used for failure on one phase only, substantial overload, short circuit, or for an earth fault; they can be locked 'out' for inspection and can be hand-reset after any fault has been rectified.

Each motor starter will normally be fitted with a 'no-volt' (or 'undervoltage') release, and three overload trips (one per phase). The overload trips may be fitted with oil dash-pot time delays which can be adjusted on site; transparent dash-pots and temperature-stable silicone fluid are preferable. As an alternative, these releases may have thermal time lags operated by the heating of bimetal strips. For low loading, overload releases may be directly connected, but for higher loading they will be operated through a transformer. No-volt releases should be arranged so that the motors will re-start automatically after a stoppage due to voltage failure.

Alarm lights, and/or bells, may be fitted to give warning of low water in a suction sump or borehole, or of overflow conditions in a receiving reservoir. These can be operated from electrodes or floats, or from level indicators; they can, if required, be arranged to operate in a building remote from the pumping station. An additional relay can be fitted so that no pump can be operated, under either manual or automatic control, when the liquid in the suction sump is below a predetermined level.

Operation of pumps is simplified by the inclusion of red and green indicating lights on each starter cubicle. A red light usually indicates that a pump is operating, while a green light will indicate that the cubicle is 'live' but that the pump is then 'stopped'. Telemetry systems for the remote control of waterworks equipment may use a different colour code, *e.g.* pump failed—red light; pump stopped normally—amber light; pump running—green light.

PLUG BOARDS, ETC.
Whether the operation of an automatic installation is controlled by floats or by electrodes, it should be possible to vary the duties of pumps. This can be arranged with a plug board or with a commutator-type sequence changeover switch so that any pump can be selected as the 'duty' pump and so that the order of the other units can be changed as required; any one of the pumps can then be selected as the standby unit. Changeover switching will normally be manually operated, but, if necessary, this can be fully automatic.

With the pneumatic bubbler type of control, a multiple switching unit takes the place of a plug board or sequence switch (*see* Fig. 7.5); this unit incorporates one switch for each pump in the installation, each switch being set so that the relay operates whenever the liquid reaches a predetermined level.

Provision for changing the sequence of pump operation will ensure that the wear on the pumps and motors is divided evenly between the various units. Maintenance will be facilitated by the provision of an 'hours run' integrator for each motor.

7 PUMP STARTERS, CONTROLS AND OTHER ACCESSORIES

HEATERS
To provide protection against the effects of condensation, automatic heating is often provided both in the starter panels and also in the windings of the motors. Provision can be arranged through auxiliary contacts for the panel heater and the heater in its associated motor to be switched on automatically when the motor is *not* in use, or these may operate thermostatically. A 'try-out' switch can be provided on the panel face so that the operation of these heaters can be checked.

EMERGENCY LIGHTING
Where a power failure is likely to occur, some precaution will be necessary when electric motors are installed. This may take the form of a second incoming supply from an alternative source, or the provision of standby diesel-driven generators or pumps.

The lighting of a pumping station will often be by electric power from a public supply, even though the pumps may be driven by other forms of power. An important station should have an emergency lighting system, supplied independently; this can either be from a generator or from a standby battery-operated supply. In a manned station, it may be preferable for an emergency system to start up automatically in case of power failure; this is important if an unexpected complete loss of lighting would be dangerous.

Emergency lighting can be provided by an automatically-started diesel-driven generator, but often the supply will be of low voltage—probably 24 volts—with just sufficient power to supply an alarm light and one or more bells, together with a few carefully sited lighting points. Such a system should be powered by a nickel-iron battery of adequate power, with its own charging equipment, and with the necessary relays to bring the lighting into operation automatically in the event of an interruption in the power supply to the normal lighting circuit.

When a low voltage emergency lighting supply is provided, it may be advisable to include a portable hand lamp to operate from a socket outlet in the starter panel. The plug for such a supply should be non-standard so that it cannot be confused with 230 volts equipment.

POWER FACTOR CORRECTION
The power factor of a circuit is obtained by dividing the watts by the volt-amperes. When this is less than unity, the power factor is said to

TABLE 7.1
MINIMUM INTERNAL RADII OF BENDS IN CABLES FOR FIXED WIRING

Insulation	Finish	Overall diameter	Minimum internal radius of bend (times overall diameter of cable)
Rubber or PVC	Non-armoured (excluding smooth aluminium sheath)	Not exceeding 10 mm	3
		Exceeding 10 mm but not exceeding 25 mm	4[a]
		Exceeding 25 mm	6[a]
	Armoured (excluding smooth aluminium sheath)	Any	6[a]
	Smooth aluminium sheath with or without armour	Not exceeding 12·5 mm	8
		Exceeding 12·5 mm but not exceeding 20 mm	10 [b]
		Exceeding 20 mm but not exceeding 30 mm	12
		Exceeding 30 mm but not exceeding 50 mm	15
		Exceeding 50 mm	18

Cable type	Sheath/armour	Overall diameter	Factor
Impregnated paper, varnished cambric, or varnished PTP fabric	Lead or corrugated aluminium sheath with or without armour	Any	12
	Smooth aluminium sheath with or without armour	Not exceeding 30 mm	12 [b]
		Exceeding 30 mm but not exceeding 50 mm	15 [b]
		Exceeding 50 mm	18 [b]
	Copper or aluminium sheath with or without PVC covering	Any	6
Mineral			6

[a] For circular conductors only. For shaped conductor cables, the minimum bending radius factor is eight. Otherwise this table applies to both circular and shaped conductor cables.
[b] The factors are applied to the diameter over the aluminium sheath.

By courtesy of AEI Cables Ltd.

be 'lagging'. A lagging power factor in an installation results in a reduction in the effective capacity of transmission and distribution lines. It is therefore usually penalised in a supply authority's tariff, and to avoid this, it may be desirable to install some form of power factor (p.f.) correction.

Synchronous induction motors have a power factor of unity (or a slight 'lead') but they are only applicable to large installations which must run for long periods; they are also only available as constant-speed motors. A squirrel-cage or a wound rotor induction motor will have a power factor of from 0·65 to 0·90, depending on its size and loading. The normal arrangement for power factor correction entails the fitting of static condensers (capacitors) across the motors, usually to ensure that the power factor will not be less than 0·95. These need not be mounted inside the starter cubicles, and it is often more practicable to mount them under the motor floor, adjacent to the motors. The capital cost of these condensers can be recovered very quickly by the saving in kVA charges. Power factor correction capacitors do not improve the efficiency of the motors; by fixing them as near to the motors as possible they do, however, improve the efficiency of both the supply undertaking's cables and the consumer's cables up to the motors.

CABLES AND WIRING

All materials and work in connection with cables and electrical wiring should comply with any local regulations. These will usually relate to the rating of cables and other equipment, and to testing, earthing and other safety aspects. Recommended bending radii of cables and the spacing of fixing clips are set out in Tables 7.1 and 7.2.

All systems which employ metal conduits and sheathing must be earthed irrespective of the voltage of the supply; earthing leads must not be insulated. Metal conduits must be electrically continuous throughout and they should preferably be of the heavy gauge 'seamless' screwed type, galvanised inside and out.

Floor-mounting starter cubicles have their cable entries at the bottom through a gland plate which is drilled on site as required. All wiring in the starter cubicle then terminates at a cable box in the base of the cubicle. Cable connections between the starter cubicles can be carried through a cable duct formed in a plinth at the base of the cubicles; connections to motors are usually taken below the motor floor and then brought up to the motors alongside the motor bases.

7 PUMP STARTERS, CONTROLS AND OTHER ACCESSORIES 79

Cable racks or trays are used to carry the cables where they are below floor level. The use of special armoured cable, or flexible sheathing, at connections to the motors improves their appearance and is a precaution against damage.

TABLE 7.2

FIXING NON-ARMOURED CABLES—SPACING OF CLIPS
(IEE Regulations Table B2)

Overall diameter of cables[a]	*Horizontal runs*	*Vertical runs*
Not exceeding 9 mm	250 mm	400 mm
Exceeding 9 mm, not exceeding 15 mm	300 mm	400 mm
Exceeding 15 mm, not exceeding 20 mm	350 mm	450 mm
Exceeding 20 mm, not exceeding 40 mm	400 mm	550 mm

[a] Major axis for flat cables.

By courtesy of British Insulated Callendar's Cables Ltd.

The load to be carried will decide the size of cable to be used. For loads up to about 150 amp, a three-core cable may be used, but for larger loads it is usually more satisfactory to employ three single-core cables.

CONTROL OF VARIABLE SPEED AND MULTI-SPEED MOTORS

Variations in the rate of output from a pump can be obtained either by variable pitch impellers or by variation in the speed of the drive motor. The use of variable pitch impellers is limited to the 'propeller' type of pump, *i.e.* axial-flow pumps; this type of pump may sometimes be used to advantage at river intakes and for land drainage schemes. The pitch of the impellers can be altered to suit variations in the required rate of pumping; this variation can be made automatically and while the plant is operating.

Variation in output of the type of centrifugal pump used for both water supply and sewerage can be obtained by variation in the speed of the motor if this is an induction motor. Speed variation can be obtained by the use of d.c. current; more usually, however, this will be by rotor-circuit resistance (with slip-ring motors) or by commutator type a.c. motors. Commutator motors have good power factors over a wide range of speeds and are not subject to 'hunting' as may happen with slip-ring motors with rotor regulation. Variation in speed is not possible with squirrel-cage motors, except by using slip couplings between the motors and the pumps.

Variation in pump output may be particularly important in a sewage pumping installation which discharges directly to a sewage treatment works. In this case there is considerable advantage in matching, as far as possible, the pump output to the sewage inflow. Fully variable speed motors are, however, expensive in capital cost and maintenance, and a more economical installation is often possible with a series of *multi-speed* motors; these can be automatically controlled.

Control of variable speed or of multi-speed motors can be by floats or electrodes in much the same way as additional constant speed motors are started and stopped. The patented form of control using only one pair of electrodes (referred to earlier in this chapter) can equally be used to automatically operate different numbers and speeds of variable speed motors.

CHAPTER 8

Ancillary Equipment

The preceding chapters have looked at various types of pumps motors and control equipment. The following chapters (Chapters 9 and 10) consider buildings and rising mains. This chapter refers to valves and pipework within the station, cranes and gantries, ladders, and other miscellaneous equipment which is required to complete the installation.

PIPEWORK

The capital cost of pipework in a large pumping station can be a major part of the total cost of the installation; the choice of suitable pipe diameters is therefore very important. Large pipes are expensive in capital cost while smaller pipes will increase the friction losses and therefore the running costs. The diameters of both suction and delivery pipes should be sufficiently large to keep velocities and head losses to a minimum (*e.g.* velocity not more than about 1·5 m/sec in the suction pipe); this may entail the use of taper pipes on both suction and delivery connections at the pump.

All internal pipework in a pumping station should have flanged joints. Each pump should have its own separate suction pipe. A common delivery main can then terminate just outside the building with a flanged socket for connection to the rising main. Suitable flexible joints should be included to facilitate dismantling and to allow for any external pipe movements. The pipework within the building to (including) the final flanged socket can be conveniently included in the pump manufacturer's contract. Wherever possible, 'special specials' should be avoided.

All pipework within the building should be supported on concrete stools or on wall brackets; stools are generally about 300 mm thick (along the axis of the pipes) and a little wider than the pipes themselves. Maximum spacings of fixings for cast-iron pipes of any diameter should be about 1·8 m on horizontal runs and 3·0 m on vertical runs. Spacings of fixings for pipes of other materials will depend on the material and on the diameter of the pipes. Safe spans for cast-iron pipes can be calculated from the formula:

$$L = 484\ D^{0.5}$$ **Formula 8.1**

where

L is the span, in millimetres,
D is the diameter, in millimetres.

All pipework and valves should be sited for reasonable ease of access for any repairs, painting, etc.

Suction pipes, when taken through a wall to a wet well, should preferably be built into the concrete when it is cast. All pipes through walls should be complete with puddle flanges, and they should have a slight rise towards the pumps to avoid the formation of air pockets. Suction pipes should terminate with a bellmouth in the wet well to reduce head loss at entry, and should preferably be turned through 90° so that the open end faces the floor of the sump. There should be a minimum clearance of $D/2$ between the bellmouth and the side wall of the sump (where D is the nominal diameter of the suction). The clearance below the bellmouth should be between D (maximum) and $D/2$ (minimum). Individual suctions should be spaced at about $4D$ between centres to avoid hydraulic interference. The lowest water level in the sump must be sufficiently far above the suction bellmouth

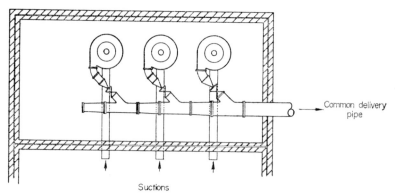

Fig. 8.1. An arrangement of pumps and rising main.

8 ANCILLARY EQUIPMENT

to avoid air being drawn in by vortex action when the pump is operating. When a common delivery main is used for a number of pumps it should be level with the outlets from the pumps to avoid vertical connecting pipes. Preferably, pump deliveries should join the common pipeline at an oblique (45°) angle. This is shown diagrammatically in Fig. 8.1.

VALVES

Sluice valves should be provided on the individual suctions and deliveries of each pump in an installation. They should preferably be installed with their spindles vertical. When fixed inside buildings they should have external screws and be opened by handwheels; these should be placed for ease of operation and should have diameters to suit the sizes of the valves. A formula often quoted is:

$$d = 6 \cdot 0(D)^{\frac{1}{2}} \qquad \text{Formula 8.2}$$

where

d is the diameter of the handwheel, in mm,
D is the nominal diameter of the valve, in mm.

With constant-speed centrifugal pumps, valves may be used to regulate the output of the pumps, but a pump should not be operated against a closed valve for more than a very short period as dangerous overheating can occur. A propeller pump or a positive displacement pump must *not* be operated against a closed valve. When a station is unattended, the delivery valve can be arranged to close automatically when the pump is stopped and to open once the pump has reached its full speed. In a sewage pumping station it is usual for the delivery and suction valve to remain open at all times when the pumps are switched to automatic control.

A reflux valve should be included on the delivery of each pump; this should be *between* the pump and the delivery sluice valve so that it can be isolated for cleaning when necessary. In a sewage pumping station the reflux valve should be fitted on a horizontal section of the delivery pipework to avoid deposits of solids between the gate and the seating. Reflux valves may be fitted in vertical pipelines in water pumping stations; they must then be of the special design suitable for such an installation.

In sewage pumping stations all valves should be of the type suitable for use with sewage and they should be complete with access doors or scour plugs. Reflux valves should preferably be fitted with external levers so that they can be opened manually to allow a backflush from the rising main to clear any solids.

VALVE HEADSTOCKS

In an installation of vertical-spindle pumps where the motors are installed on a floor above the pumps it is general for all valves (suction and delivery and any auxiliary suctions) to be operated from the motor floor. This ensures that operation of the valves is possible at all times (including any flooding of the pump basement) and also that the valves are opened whenever any pumps are switched on.

The valve spindles can be extended, through holes in the motor floor, and connected to headstocks fitted to the floor. Handwheels are fixed either immediately above the headstocks, or to one side (through gearing); when required, indicators can be fitted to show the relative positions of the valve gates. Where the extension spindles are long, intermediate guide brackets will be required. Motorised opening may be necessary for large valves.

To present a neat appearance at motor floor level, the sets of valve headstocks should be lined up parallel to the motors; they should be arranged so that it is obvious which valves relate to each motor. Where necessary, headstocks can be offset from the valve spindle centreline by using suitable universal joints in the spindle extensions.

When headstocks are fitted outside a building (*e.g.* over a wet well of a pumping station) they should be locked to prevent unauthorised operation.

AIR RELEASE COCKS

Care must be taken in the siting of pump suctions to prevent the ingress of air, as air entrained in the liquid will adversely affect the performance of a centrifugal pump. Pumps should be fitted with air release cocks to operate automatically to release, from the pump casing, air which would otherwise be trapped while the pump is starting. In sewage pumping stations they should be fitted with extension pipes so that they will discharge into the wet well or outside the station.

The air release cocks should be fitted at the highest points of the pump casings and, if on sewage pumps, they should be suitable for

8 ANCILLARY EQUIPMENT

use with liquids containing solids. An isolating valve should be fitted below each air release cock to allow the cock and the pipeline to be serviced from time to time. The pipeline should be of 20 or 25 mm diameter.

PRIMING DEVICES

The air release cocks referred to above will allow centrifugal pumps to be self-priming provided they are located in a dry well alongside the liquid in the wet well. If the pumps are located above the starting water level, either foot valves must be fitted (so that the pumps can be primed by hand or from the rising main), or an air extractor pump must be fitted. In any event, centrifugal pumps are rarely installed at more than three metres above starting water level.

In sewage pumping stations both foot valves and extractor pumps can be affected by the solids in the liquid, and wherever possible the pumps should be located below the starting liquid level, so that their use can be avoided. Automatically controlled priming is an added complication in any pumping station, and is yet another item of equipment which may fail to operate when required. This is one of the main reasons for the more general use of vertical-spindle pumps for automatic sewage pumping stations.

In a 'self-priming' centrifugal pump, the impeller is so designed that it traps a certain amount of liquid to enable it to start the pumping action; the air must then be exhausted from the pump casing at a regulated rate to maintain the pumping action. One form of self-priming pump for sewage duties has a priming vessel set above the pump itself. Reciprocating pumps are self-priming, with a suction lift, provided the glands are airtight and that the pump casing clearances are not excessive; otherwise priming arrangements are required in the same way as for centrifugal pumps.

PRESSURE AND VACUUM GAUGES

A pressure gauge should be fitted to the delivery pipe of each pump; the gauge dials (suitably corrected for the difference in height) can be mounted in the motor room. A vacuum gauge should be fitted to each suction pipe as near to the level of that pipe as possible. Connecting pipes to the gauges should be of heavy-gauge copper or nylon tube, and should be complete with two-way cocks, so that they can be cleaned of any obstruction without the need for dismantling.

Best quality chromium plated gauges are generally used, and their dials should preferably be at least 150 mm diameter. If the pressure gauges can be grouped on to one polished wooden board in the motor room, they will present a tidy, orderly appearance and they can be conveniently referred to when the pumps are operating.

SCREENS AND COMMINUTORS

As far as possible, an automatic sewage pumping station should be *fully* automatic, *i.e.* it should be possible to leave the pumps unattended for fairly long periods. The pumps should therefore be capable of handling crude unscreened sewage so that screening and grit removal can be carried out later at the sewage treatment works. A coarse screen will, however, often be considered necessary to protect the pumps. In some circumstances 'medium' screening (bars spaced at 15 to 20 mm) or comminution may be installed. When appropriate, the screens can be mechanically raked. The screens or comminutors should be located in a special chamber at the inlet to the station.

The disposal of screenings at an isolated sewage pumping station is always a problem, and they will be a source of smell nuisance. In a larger pumping station it is sometimes possible to return the screenings through a disintegrator to the main sewage flow; for a smaller pumping station a comminutor may be more suitable.

For a relatively small water supply installation, a combined footvalve and strainer may be sufficient protection for a pump. For larger water supply intakes at rivers, bar screens and self-cleansing fine screens will usually be installed. Coarse screens may be fixed to exclude weeds, reeds and other floating debris, and these will then be followed by the fine screens (often band, disc or drum screens).

GLAND DRAINAGE

Attempts should not be made to stop leakage from glands by the excessive tightening of gland nuts, as slight visible leakage should always be present. It is usual to drain the dry well floor towards a sump (floor slope of about 1 in 100 minimum), which can be emptied either by a special sump pump, or (in sewage installations) by auxiliary suctions. These have been referred to earlier in Chapter 5.

Auxiliary suctions are small suction pipes of about 30 or 40 mm diameter, connected into the main suction pipe of a sewage pump (upstream of the sluice valve) and operated by a full-way (plug cock) valve. A sump pump is generally preferred; this should operate automatically, controlled by the level of the water in the sump. The pump room floor can be kept dry by fitting small diameter drain pipes from the glands of each pump to discharge into the sump.

TOOLS AND SPARES

On commissioning a pumping station, provision should be made for the supply of a set of tools for the pumps and motors, and also a reasonable supply of spare parts. Spares to be stocked will depend on the location of the pumping station; an installation overseas or in a remote part of the country should carry a wider range of spares than installations nearer the manufacturer's works. The minimum spares should include starter panel contacts, fuses, pump packings, and probably pump bushes and glands. The inclusion of a strongly bound log book at this stage should ensure that a log is maintained from the initial commissioning.

The Specification (*see* Chapter 11) will normally state that a set of polished spanners (mounted on a polished hardwood board for wall mounting) are to be provided; they should not be used during installation but should be handed over in a new condition. The Specification may quote certain spares and lubricants to be provided with the equipment, but it is generally preferable to invite the manufacturer to include, in his tender, for any extras which he may consider necessary. The spares should be provided in a suitable box which will normally be kept at the pumping station.

A complete set of wiring diagrams for all electrical equipment should be provided on suitable durable material and fixed to the wall of the motor room for easy reference. Sets of characteristic curves should be provided by the pump manufacturer when the equipment is installed; these are best carefully filed at the headquarters office.

LADDERS AND STAIRWAYS

Access to the basement of a very small pumping station may sometimes be by a vertical ladder; a set of stairs with non-slip treads, set at a reasonable pitch and equipped with handrailing will, however, be safer and will do much to encourage regular maintenance. Straight run stairways are preferable to circular (spiral) stairways. When a station substructure is very deep (more than about 6 m), an intermediate platform will be necessary whether access is by stairs or by ladders. Vertical ladders should be complete with safety cages and

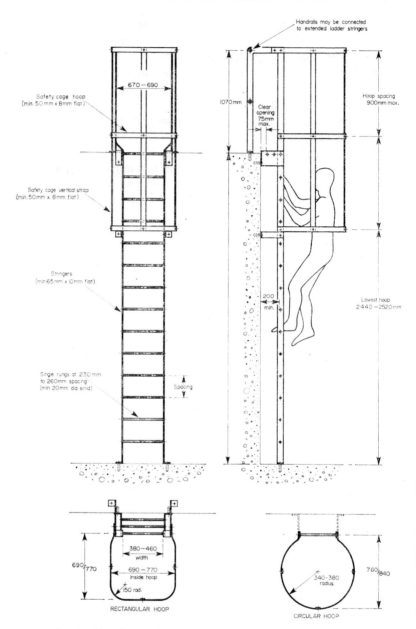

Fig. 8.2. *Ladder details* (by courtesy of Steelway, Glynwed Tubes and Structures Ltd.).

extension stringers carried up above the upper floor level (*see* Fig. 8.2).

For sewage pumping stations, a vertical ladder into the *wet* well is generally considered to be satisfactory; access to this should be from the open air. Access to the *dry* well should preferably be by a stairway from inside the motor room. The best design of stairway for ease of use is based on a figure of from 600 to 630 mm for 'twice the riser plus the tread' ($2h + d$), where d is 250 mm *minimum* and h is 190 mm *maximum*.

HANDRAILING

As stressed in 'Safety in Sewers and at Sewage Works' [34], handrailing should be provided where any danger exists. Handrailing should be fixed around all openings in pumping station floors (including the stairwell), with gates or removable sections of chains where required. This includes locations where there is any difference in level or where there is any other danger of an operator falling. The handrailing can either be of heavy duty steel or aluminium and is generally of the two-rail type with the upper rail set about 1 m above floor level. Steel handrailing tubing and standards can be painted, galvanised, stove-enamelled, polished, chromium-plated, or plastic covered. Plastic coating gives a clean hygienic finish which does not need painting. Handrailing of aluminium alloy needs little or no maintenance. Chains for closing off openings can be black, galvanised or polished finish, and (to match plastic-coated handrailing) bright chains can be obtained with plastic sheaths.

Handrailing standards can be obtained for fixing to the top of concrete floors or to steel joists, for side fixing, for building into floors, or for slotting into specially provided sockets. Special standards with angled bases can be obtained for use with staircases.

FLOOR PLATES

When vertical-spindle pumps are installed, openings must be left in the structural floor so that the pumps and other equipment can be lifted out for repairs or replacement. Removable sections of flooring should be of steel or aluminium chequer plating, cut to sizes and shapes that can easily be lifted, and set into angle frames which provide a stop for any motor floor finishes (quarry tiles, etc.). Open-type flooring can be used for stairways and intermediate platforms, but is generally not considered suitable for use at motor floor level, as

TABLE 8.1
ADMIRALTY PATTERN CHEQUER PLATES, DIMENSIONS AND WEIGHTS

Thickness on plain (mm)	Maximum length (m)	Maximum width (m)	Maximum area (m^2)	Approximate weight (kg/m^2)
4·76	10·67	1·92	19·51	42·53
6·35	12·19	2·13	23·23	54·98
7·94	12·19	2·13	23·23	67·43
9·53	12·19	2·13	23.23	79·88
11·11	12·19	1·98	22·76	92·33
12.70	12·19	1·83	20·90	104·78
14·29	12·19	1·68	19·51	117·23
15·88	12·19	1·52	16·72	129·86

By courtesy of the British Steel Corporation (Northern and Tubes Group)

a spanner or bolt dropped through the flooring can injure a man below, or can damage machinery. Weights and safe loadings of floor plates are given in Tables 8.1 to 8.3. Where there is a possibility of machinery being temporarily placed on flooring panels they must be strong enough to take the concentrated loading; in some cases this may be classed as a 'live load'.

TABLE 8.2
ADMIRALTY AND DURBAR PATTERN STEEL PLATES: SAFE LOADINGS—SIMPLY SUPPORTED ON TWO SIDES (kg/m^2)

Thickness on plain (mm)	Span (m)							
	0·6	0·8	1·0	1·2	1·4	1·6	1·8	2·0
4·76	1 389	781	500	347	255	195	154	125
6·35	2 470	1 389	889	617	454	347	274	222
7·94	3 859	2 170	1 389	965	709	543	429	347
9·53	5 556	3 125	2 000	1 389	1 021	781	617	500
11·11	7 563	4 253	2 723	1 891	1 389	1 064	840	681
12·70	9 878	5 556	3 556	2 470	1 814	1 389	1 098	889
14·29	12 502	7 031	4 501	3 126	2 296	1 758	1 389	1 125
15·88	15 435	8 681	5 556	3 859	2 835	2 171	1 751	1 389

By courtesy of the British Steel Corporation (Northern and Tubes Group)

8 ANCILLARY EQUIPMENT

CRANES AND GANTRIES

Pumping plant requires maintenance from time to time to replace impellers and other parts; this involves the lifting of the motors and shafting, and sometimes the pump units themselves. Except for very small stations, overhead lifting gear should be provided for the installation or the removal of pumps and other plant. This may be a travelling gantry crane or a travelling hoist on a fixed runway beam.

For a small station it may be sufficient to fix a simple joist over the centre lines of the motors or pumps; this can then carry a chain block or a portable electric hoist when needed. Alternatively, a portable gantry may be used; models are available which can be dismantled and taken through an ordinary doorway. Except in very large stations, motorised crane traversing is rarely justified. With the advent of large mobile cranes, there may be some future in the provision of removable roof sections to the pumphouse; this would allow whole units to be removed for servicing at a central workshop.

Sufficient headroom must be allowed in the motor room to house the lifting tackle and a unit of equipment under the gantry, and still leave clearance over other motors and starter cubicles, etc. so that the unit to be serviced can be moved as required. Adequate steel cable must also be available, so that the crane hook can be lowered to the bottom of the dry well. The strength of the gantry or lifting beam must be sufficient for the heaviest load likely to be lifted; this is generally the largest electric motor installed in the station.

HEATING AND LIGHTING

In many pumping stations the provision of both heating and lighting will be by electric power; in some circumstances external heating may be available to heat hot water radiators. Some form of heating should be provided in any station (even an unmanned station) to avoid damage to machinery and to provide suitable conditions for working, either routine or for emergency repairs. Automatic thermostatic control is preferable to prevent freezing in cold weather. This heating is generally considered to be necessary *in addition to* any heaters fixed in the motors and starters (*see* Chapter 7). Some space heating will of course be available from the heat generated by the motors (including electric motors).

The lighting of a station should be adequate for all normal maintenance, and its siting and intensity should be sufficient for all normal work in the station; hand lamps should only be required for

TABLE 8.3
ADMIRALTY AND DURBAR PATTERN STEEL PLATES: SAFE LOADINGS—SIMPLY SUPPORTED ON FOUR SIDES (kg/m^2)

Thickness on plain (mm)	Span (m)									Breadth (m)
	0·6	0·8	1·0	1·2	1·4	1·6	1·8	2·0		
4·76	2 778	1 828	1 570	1 476	1 437	1 416	1 406	1 400		0·6
		1 563	1 102	936	864	830	812	801		0·8
			1 000	741	631	576	548	531		1·0
6·35	4 939	3 250	2 790	2 625	2 554	2 517	2 500	2 489		0·6
		2 778	1 959	1 664	1 537	1 476	1 444	1 425		0·8
			1 778	1 317	1 121	1 024	974	945		1·0
				1 235	952	812	739	698		1·2
7·94	7 717	5 077	4 360	4 100	3 990	3 933	3 905	3 890		0·6
		4 341	3 061	2 599	2 401	2 306	2 256	2 226		0·8
			2 778	2 058	1 752	1 600	1 521	1 476		1·0
				1 929	1 487	1 269	1 155	1 090		1·2
					1 417	1 125	968	879		1·4
9·53	11 113	7 311	6 278	5 905	5 746	5 664	5 624	5 601		0·6
		6 251	4 408	3 743	3 457	3 321	3 249	3 205		0·8
			4 000	2 963	2 522	2 304	2 191	2 126		1·0
				2 778	2 142	1 828	1 663	1 569		1·2
					2 041	1 620	1 394	1 266		1·4

8 ANCILLARY EQUIPMENT

11·11	15 125				7 821	7 709	7 655	7 624	0·6
		9 951	8 545	8 037	4 706	4 521	4 422	4 363	0·8
		8 508	6 000	5 094	3 433	3 136	2 982	2 893	1·0
			5 445	4 033	2 915	2 488	2 264	2 136	1·2
				3 781	2 778	2 205	1 898	1 723	1·4
						2 127	1 726	1 500	1·6
12·70	19 756				10 215	10 069	9 998	9 957	0·6
		12 997	11 161	10 497	6 146	5 904	5 776	5 699	0·8
		11 112	7 837	6 654	4 484	4 097	3 895	3 779	1·0
			7 112	5 268	3 807	3 249	2 957	2 790	1·2
				4 939	3 629	2 880	2 478	2 251	1·4
						2 778	2 255	1 959	1·6
14·29	25 003				12 928	12 744	12 653	12 602	0·6
		16 449	14 126	13 285	7 779	7 473	7 310	7 212	0·8
		14 064	9 918	8 421	5 675	5 185	4 929	4 783	1·0
			9 001	6 667	4 819	4 112	3 743	3 531	1·2
				6 250	4 592	3 645	3 137	2 849	1·4
						3 516	2 854	2 480	1·6
15·88	30 868				15 961	15 733	15 622	15 559	0·6
		20 308	17 440	16 402	9 603	9 226	9 024	8 904	0·8
		17 363	12 245	10 397	7 006	6 401	6 085	5 904	1·0
			11 112	8 231	5 949	5 077	4 621	4 360	1·2
				7 717	5 670	4 499	3 873	3 517	1·4
						4 341	3 523	3 061	1·6

By courtesy of the British Steel Corporation (Northern and Tubes Group)

close inspection. When required, additional lighting should be possible from suitably sited wall sockets. For large manned stations, as much use as possible should be made of natural lighting from windows and roof lights. For unmanned stations, and especially for isolated buildings, it may be advisable to *reduce* the area of windows to conserve heat and to combat vandalism; more reliance will then be placed on artificial lighting. Fluorescent lighting should be of a type which will avoid stroboscopic effects. The lighting of substructures should be controlled from switches at the motor room floor level.

Where a composite switchgear and starter panel is to be installed, it may be convenient to arrange for the heating and lighting sub-circuits to be taken off this panel and for their isolating switches and fuses to be included in the panel. Lighting circuits in particular must be connected to the 'live' side of the main switchgear circuit breaker so that lighting will be available for repairs and maintenance at the panel.

VENTILATION

Natural ventilation through doors, windows and roof ventilators is generally adequate for smaller pumping stations. When a motor room is to be constructed underground, mechanical ventilation may be required. As a general rule, the quantity of air required to dissipate the heat generated from electric motors will be about 10 m^3 per kW of total power operating at any one time, assuming a rise in air temperature of about 5°C.

At depths below about 10 m, forced ventilation may be required to remove fumes and gases from the dry wells of larger pumping stations. Ventilation of the wet well of a sewage pumping station will also assist in the ventilation of the sewers discharging to it. Vent pipes provided for this purpose must not be used for other purposes, *e.g.* rainwater downpipes.

AMENITY AND SAFETY

The provision of basic amenities for operating staffs at pumping stations is frequently overlooked. Except in the smallest of unmanned stations, the minimum provision at any pumping station should include washing and toilet facilities. A supply of clean water is even more important at a sewage pumping station, both for the operator and to supply a hose for washing down screens and benchings in the wet well. At a water pumping station a septic tank will be necessary

8 ANCILLARY EQUIPMENT

if the toilet cannot be connected to a public sewer; care must then be exercised in the disposal of the effluent.

If any details of flows, etc. are to be recorded by the operator, a simple table and chair are almost indispensable. For a larger station, the facilities to be provided will depend on the number of staff and the length of time they are employed at the station each day; it may be relevant to provide a messroom and drying rooms, and even a small office. Consideration should also be given to space for the parking of cars and bicycles.

All openings in floors should be provided with either handrailing or cover plates. Rotating shafts, chain drives, etc. must be protected with guards; rubber mats should be provided for use in front of electrical switchgear panels, especially if these are not of the enclosed type.

If a public water supply is to be used to provide a supply of clean water at pump stuffing box seals, there must be no possibility of pollution from a back flow into the water system. This will then normally entail pumping the gland water to obtain the necessary pressure.

In the interests of safety it may be possible to have a telephone installed in an isolated pumping station; this is, of course, less important when the operating gangs have radio contact with their headquarters.

CHAPTER 9

Buildings

A pumping station building is probably one of the few components of a water or sewerage scheme ultimately left visible to the general public. Its cost will often be only a very small part of the full cost of a scheme but it will remain for ever an example of the designer's skill and also of the standard of maintenance of the authority. A little extra spent on the pumping station building and its surroundings is generally well worth while as a well designed and operated station will become a part of the landscape. Poor design and, particularly, poor maintenance can easily result in complaints of smell and noise. Poor standards of painting, cleanliness and repairs at a pumping station so often reflect general poor standards of the authority concerned.

The approximate site of a pumping station is set, to a great extent, by technical considerations, but its final detailed siting can often take into account the amenity of the locality; an adjustment of (say) 50 m at this stage may be of considerable importance to some nearby residents. In some circumstances a completely underground station may be the answer.

The site must be free from the liability to flooding and must be readily accessible for both construction and maintenance. To avoid damage from sun and driving rain it is preferable to have the main doorway facing East or North, if possible.

While this chapter refers to pumping station buildings generally, for both water supply and sewerage schemes, it particularly relates to buildings for automatic sewage pumping stations.

EMERGENCY OVERFLOW

Some sewage pumping stations are provided with overflows. If the sewers are on the 'combined' system, storm sewage overflows may be

9 BUILDINGS

included within the system; in some cases these may be sited at pumping stations where more positive control may then be possible. While emergency sewage overflows are not generally welcomed by river authorities, they are a wise precaution where there is the slightest possibility of pump failure. This will always be a possibility with automatic electrically-operated pumps. This type of overflow should be set as high as possible in the wall of the pumping sump (wet well) so that its operation is delayed as long as possible. The provision of an overflow presupposes the availability, near the pumping station, of a suitable watercourse, and one should not be permitted where there is any hazard to health or possibility of damage to private property. Alarms to give warning of non-operation of pumps were referred to in Chapter 7.

THE WET WELL

The conventional sewage pumping station houses vertical-spindle pumping units with the pumps located in a dry well, drawing sewage from an adjacent wet well. The capacity of this wet well will depend on the number and capacity of the pumps and on the design of the starters. The calculation of storage capacity between 'start' and 'stop' levels has already been referred to in Chapter 7.

The depth of the well will be set by the depth of the incoming sewer; to avoid silting in the sewer due to low velocities, the highest 'cut-in' level of the normal duty pumps should be below the normal dry weather flow depth in that sewer. In fact, the cut-in level is often set at 150 to 300 mm below the sewer *invert*. The 'cut-out' (or 'stop') level of a pump is usually set at about 400 to 500 mm below the start level. Each pump must be located with its gland below the 'start' level so that the pump is fully primed before starting; the importance of air release valves on the pumps has been referred to in Chapter 8. The location of the pumps and their suction pipework will then determine the depth of the wet well below sewer invert level (*see* Fig. 4.2).

The wet well usually extends the full length of the dry well of the pumping station, and should be divided into two compartments for easy maintenance. The partition wall should have an isolating penstock so that any one or both compartments can be used as required. The length of the wet well will therefore depend on the number of pumps and their spacing, pumps normally being at between two and three metres centres to allow room for maintenance.

The incoming sewer should discharge through an inlet chamber to either or both compartments. Cascading and turbulence should be avoided; particularly in the vicinity of the suction pipes and also near any electrodes or floats. The floor of the wet well should slope at 1 in 1, or steeper, to a sump at the pump suctions. Many engineers prefer a slope of about 1·75 vertical to 1 horizontal, but on the other hand these steeper slopes are inconvenient for maintenance. It may be preferable to construct a long, deep channel at the suction pipes, with a benching sloping at about 1 in 6; if this form of construction is used the lowest pump cut-off level should then be in the channel itself (*i.e.* below the level of the benching).

The capacity required between 'start' and 'stop' levels will be based on Formula 7.1. Allowing for the 400 to 500 mm depth referred to above, it is now possible to calculate the width of the wet well, allowance being made for any reduction in cross-section due to the sloping sides.

Access to the wet well should be by manhole covers from outside the pumping station building; if two covers are provided, through ventilation can be arranged while men are working in the well. The access should be large enough to facilitate cleaning down inside and for the withdrawal of any suction piping. Vent pipes must also be provided for normal ventilation—these pipes should be taken up to above the eaves of the building. When the wet well is wide enough, it is generally preferable to arrange for the motor room floor to extend over a part of this, allowing any level electrodes to be brought up through the floor to the motor room where they will be protected from the weather and vandals (*see* Fig. 4.1).

Any lighting in the wet well must be located above top water level and all fittings and cables should be weatherproof and flameproof. For very small stations, sufficient lighting in the wet well can probably be obtained by using a portable inspection lamp; this should, however, be of the gasproof and flameproof variety.

The substructure of a pumping station should be designed as a water retaining structure. Mass concrete or brickwork structures are rare now, and both wet and dry wells will usually be of reinforced concrete. Sewage pumping stations are generally sited at low points and in these circumstances precautions are generally required to prevent flotation when the wells are empty. This precaution is particularly important *during construction, i.e.* before the wet well is filled and before any machinery or superstructure is in position. The walls of the wet well should be floated to a smooth finish. Ledges and platforms that might collect solids should be avoided.

The construction of pumping stations with circular substructures has been found to be convenient in soft or waterbearing strata. These can be sunk under their own weight, concrete being poured as the ring is sunk, and excavation being carried out from *inside* the ring as necessary. Hood [7] has reported that in Sydney (Australia) some smaller pumping station wet wells have been formed of two precast concrete tubes, set vertically side by side, with an interconnecting pipe and sluice valve. Suction pipes are then taken to these rings from the pumps which are installed in a further cylinder; this dry well cylinder may then be about 3·5 or 4·0 m diameter for a small station.

THE DRY WELL

To ensure that the pumps are primed at all times (particularly for automatic operation) the pumping station layout often incorporates an underground 'dry well' alongside the wet well. This dry well can house either horizontal-or vertical-spindle pumps for either water supply or sewerage. The use of underground wet and dry wells is particularly common for automatic sewage pumping stations (*see* Fig. 4.2). The dry well is not required if submersible pumps are used; in this case the pumps will be installed in the wet well (Figs. 4.1 and 5.1).

The dry well (or pump chamber) should extend to at least as low as the bottom of the suction sumps (in the wet well); the pumps can then be set as low as possible to take full advantage of the liquid level to keep the pumps primed. In some circumstances the motors can be fitted immediately above vertical-spindle pumps, so dispensing with the need for a motor floor and a relatively high building; provision must then be made to house the control gear. The design of the building is, however, simplified by the inclusion of a motor floor at ground level, and for sewage pumping stations this is by far the most usual form of design. Where the station is very deep it may be convenient to include an intermediate floor or platform to avoid long stair runs; if a floor is included, suitable openings must be incorporated to allow the installation and removal of pumps, pipes, etc.

The size of the dry well is dictated by the number and size of the pumps to be accommodated, together with the space required between each pump and between the pumps and the walls; this space is needed so that maintenance can be carried out in safety. Generally a minimum clearance of one metre will be required, together with further space for reflux and sluice valves, sump pump, etc., and where relevant, sufficient space to carry out minor running repairs (including possibly a small work bench). The spacing of pumping units has already been referred to in Chapter 8, and consideration must always be given to possible future additional units.

The floor of the dry well should slope gently towards a channel leading to a sump; this sump will house a sump pump or the auxiliary suctions from the pumps (*see* Chapter 8). The pumps will normally be set on concrete plinths on the dry well floor. These plinths and any supporting blocks for pipework and valves must be securely fixed to the floor; it is usual to extend the reinforcement from the floor up into the plinths.

The floor should be finished smooth for ease of maintenance; in some cases (particularly in water supply pumping stations) the floor may be tiled. Any flooring finish must, of course, be resistant to oil. The walls should be finished clean and smooth; this may entail rubbing them down with carborundum. They should generally be given one or two coats of light coloured paint. Prefabricated factory-built pump chambers are available; these are discussed in more detail below.

Access to the dry well should be by a stairway, except in very small stations where a ladder may suffice. This stairway must be carefully sited to fit in with the equipment and pipework, and should be complete with all normal handrailing. Chapter 8 contains more information on stairways, etc. and also on floor plates used to cover openings left in the roof of the dry well. This roof will usually form the floor of the motor room; it must be sufficiently strong to carry all possible loadings and should be finished smooth and clean on its underside. Various types of floor using hollow blocks or concrete plates, and also special shuttering systems of steel plates or polypropylene moulds, are available which can simplify and speed up the construction of suspended floors.

Lighting fittings below the motor floor should be weatherproof and flameproof, and they should be controlled from the motor floor; the lighting in the dry well should be adequate for safe working without the normal need for portable lamps. Good lighting encourages good maintenance.

THE SUPERSTRUCTURE

While the size and height of the motor room will be dictated by the technical requirements of motors and cranes, the aesthetic design of the overall building must take into account the amenity of the locality. The style and finish of the building should be chosen depending on whether the pumping station is in a village or in the centre of a town. In any case the appearance should be modern and attractive, and a stereotyped pattern should be avoided. The superstructure will generally be of brickwork, with either a pitched or a flat roof; the facing bricks and roofing tiles should be of high standard and should, if possible, blend with any neighbouring buildings. In a clean, well-designed building the maintenance of the machinery will usually be of a high standard.

TABLE 9.1

COMPOSITION OF MORTARS FOR BRICKWORK

Mortar class	Proportions						Min. compressive strength required (N/mm^2)	
	Volume			Weight				
	Cement	Lime	Sand	Cement	Lime	Sand	7 days	28 days
A				1	0	2^a		24·1
B	1	0·25	3	1	0·1	4	11	16·6
C	1	0·5	4·5	1	0·3	6	5·5	8·3
D	1	1	6	1	0·45	8	2·8	4·1
E	1	2	9	1	0·9	12	1·4	2·1
	Masonry cement			Masonry cement				
F	1		4·5	1		6	2·8	4·1
G	1		6	1		8	1·0	1·4

^{a} Or a mix of sand with epoxy resin, etc.

Notes: 1. With certain sands adequate workability can be gained with a reduced lime content, possibly with the addition of a plasticiser.
2. Compressive strengths specified are for cubes made in the laboratory. Cubes made on site should have strengths not less than 2/3 of those specified.

By courtesy of the Ibstock Brick and Tile Co. Ltd.

Various aspects of brickwork details, mortars, damp courses, etc. are referred to in the author's book on *Wastewater Treatment* [26]. Tables 9.1 and 9.2 are reproduced from that book. The external faces of the motor room walls will often be finished in fair-faced brickwork;

TABLE 9.2

RECOMMENDED MORTAR MIXES FOR VARIOUS CONDITIONS

Construction	Degree of exposure	Time of construction[a]	Recommended mortar class (see Table 9.1)	
			Ceramic bricks	Calcium silicate bricks
Copings	all	winter	B	C
		summer	BC	C
Parapets	all	winter	BC	C
		summer	BC	C
Retaining walls	all	winter	B	C
		summer	BC	C
Freestanding walls	all	winter	BC	C
		summer	BCD	C
Manholes	—	winter	BC	BC
		summer	BCD	BCD
Work below d.p.c.	all	winter	BC	CD
		summer	BCD	CD
Reinforced brickwork	all	winter	AB	ABC
		summer	AB	ABC
2-storey housing	sheltered	winter	DEF	EF
		summer	DEF	F
Low rise internal work	—	winter	BCDF	CDF
		summer	BCDEFG	CDEFG

[a] As a *general* guide, summer is April to November, winter is December to March.

By courtesy of the Ibstock Brick and Tile Co. Ltd.

internally they may be finished with plaster, plasterboard or tiles. Good fair-faced brickwork surfaces are rarely possible on internal faces due to variations in brick dimensions. External walls can have a rendered finish if required but a good quality facing brickwork is generally preferable.

9 BUILDINGS

The building should be designed to use the recommended 300 mm module, both horizontally and vertically, and to make as much use as possible of standard sizes of doors, windows, timber, etc.

The need to heat pumping stations to prevent damage to machinery has been referred to earlier. Freezing must be prevented, particularly at tubing to pressure gauges, etc.; a minimum temperature of about 10°C is desirable. Loss of heat can be prevented by suitable insulation of the building and of the roof space, but care must be taken to see that the building does not overheat in the summer.

Some pumps with weatherproof types of motors and starters can be installed in the open with no building at all; these are however comparatively rare for either water supply or sewerage schemes.

UNDERGROUND AND PREFABRICATED STRUCTURES

In some situations the pumps can be installed in an underground station if a building would not be appropriate. With most underground installations the only equipment above ground will be a small switchgear cabinet; this can often be sited inconspicuously against an existing wall or building.

For smaller sewage pumping installations the use of submersible pumps in an underground chamber will show a considerable saving in cost. Figure 5.1 shows typical chambers for one or two pumps. These pumps are reliable in use and when installed within the boundary of the public highway, the cost of land acquisition is also avoided. Other installations needing no superstructure include the SPP Solids Diverter and various types of packaged units which are supplied complete with their own factory-built chambers.

Many water boreholes are now equipped with submersible pumps; no building is then required for motors or pumps, and the switchgear can often be located away from the borehole. Submersible pumps are also used for boosting and can be installed in underground chambers with only small cubicles above ground to house the control gear.

WINDOWS: LIGHTING AND VENTILATION

In industrial buildings, considerations of economy frequently dictate the area of windows to be provided; conservation of heating often being more important than the cost of artificial lighting. To a great extent, the same considerations apply to an unmanned pumping station for either water supply or sewerage. A building *without* windows is not an attractive proposition, but where these can be limited to small openings set high in the walls, the risk of damage by vandals is considerably reduced, and any nuisance due to noise will also be minimised. In a manned station the risk of wilful damage is reduced, and working conditions are generally improved by the provision of larger windows set lower in the walls—about 15% of the floor area is then generally considered to be a satisfactory area for the windows. Artificial lighting and ventilation were referred to in Chapter 8.

Whenever windows are provided, consideration must be given to the access to them for cleaning, repairs and painting. Wherever possible, access should be available without the use of special ladders. The use of toughened glass (minimum recommended thickness, 5 mm) will reduce the possibility of damage. Special glass- or plastic-filled reinforced-concrete blocks may be suitable for use either in the walls of a superstructure or as 'pavement lights' over a wet well. Unreinforced glass blocks are non-load-bearing, but will carry their own weight and can be used, provided suitable framing supports are incorporated.

Forced ventilation is generally not required at a pumping station, and when electric motors are installed, natural ventilation by windows and/or roof ventilators is usually sufficient for both the dry well and the superstructure; air bricks can also be incorporated at ceiling level. It is sometimes advisable to install extractor fans to withdraw air from the bottoms of dry wells, particularly if they are deep. When oil-engines are in use on continuous duty, more ventilation will be required; the air required by the engines themselves should also be drawn from outside the building. In the tropics, both ventilation and the siting of windows require special attention in relation to the sun and driving rain (*i.e.* the prevailing wind).

ACCESS TO THE BUILDING

Road access is required originally to most pumping stations to facilitate the installation of the machinery; this should then be available eventually for maintenance and replacements. Access will also be required to the sump of a sewage pumping station to allow the removal of grit and screenings from time to time. Any road should be wide enough to take a mobile crane (if one is likely to be used), and it should give access to the main doors of the building. A minimum width of 3·5 m is recommended. Sufficient room should generally be available for vehicles to turn round within the site.

9 BUILDINGS 105

Roads should be well drained and constructed of materials that do not need too much maintenance. The choice between concrete or tarmacadam for road surfacing is often a matter of personal preference. Concrete roads are more easily kept clean and require very little maintenance. Tarmacadam roads need attention from time to time as potholes appear, and they should be surface-dressed with emulsion or hot tar at least every four or five years. Adequate provision must be made for the drainage of surface water (this can sometimes be taken into the wet well of a sewage pumping station); the spacing of gullies and the design of surface water sewers is considered in *Public Health Engineering—Design in Metric: Sewerage* [25].

FENCING
Fencing is an expensive item to maintain but it should be provided around a pumping station site to protect the equipment from damage and also to minimise the dangers to unauthorised persons coming into contact with electrical transformers and switchgear. Where an access road is provided (at all except the smallest of stations), a pair of double gates will be required. A hedge may sometimes be planted along with the fence to improve the appearance.

Fences should be strong and unclimbable but, if possible, they should also be attractive. They should have a long life and require the minimum of maintenance. Unclimbable fences of wrought iron or steel, up to 1·8 m or more in height, have been used for pumping stations for many years; they must be painted periodically to prevent corrosion. Chain-link fences are generally cheaper but are more easily damaged. These can be of galvanised mild steel but more generally they now have a PVC coating over either a plain or a galvanised steel core. Chain link fencing is available in heights from 0·9 to 3·6 m, in 25, 38, 44 and 50 mm mesh, and in various wire gauges. A suitable fence for a pumping station site would probably have a 50 mm mesh and be either 1·8 or 2·1 m in height, fixed on concrete posts.

LANDSCAPING
The area chosen for a pumping station site should be adequate to allow for proper treatment and landscaping of the area around the building once construction has been completed. Suitable landscaping is essential if the building is to blend with its surroundings; this

becomes particularly important in an urban area. Any treatment of the site should not be 'fussy' however, and should generally be as simple as possible to reduce maintenance costs. Steeply sloping banks should be avoided if possible; gentle slopes are much easier and less expensive to maintain. Slopes should generally be no steeper than about 1 in 2·5.

Suitable mixes of grass seed for both flat areas and banks are given in Table 9.3. Grass seed can be sown at between about 30 and 50 g/m^2. The appearance of larger grassed areas can sometimes be improved

TABLE 9.3

GRASS SEED MIXTURES

Type of grass	Flat areas (%)	Banks (%)
Poa Annua	30	50
Agrostis Stolonifera	—	20
Browntop	—	15
Crested Dogstail	30	—
Chewings Fescue	10	—
Creeping Red Fescue	10	15
S 23 Short Ryegrass	20	—

by the planting of shrubs and ornamental trees, provided their roots will not affect any pipelines. Suitable quick-growing flowering shrubs include *Buddleia, Philadelphus, Berberis, Weigelia* and *Forsythia;* among the flowering trees are the almonds, cherries and other prunus varieties. For a more rural setting, larger 'forest type' trees are preferable; these include the silver birch and some types of willow.

CHAPTER 10

Pumping Mains

A book on pumping stations would not be complete without some reference to pumping mains (rising mains). While some land drainage pumping stations discharge directly to open channels at the higher level, the majority of pumps for water supply and sewerage are arranged to discharge through pipelines. The design of a pumping main must take into account the head losses imposed by friction, while the choice between different diameters and materials will be partly a matter of economics. Velocities of flow in the rising main will be important (particularly for sewage pumping).

A rising main will normally follow the contours of the ground, but the individual gradients of sections of the pipeline will not influence its design. The carrying capacity of a rising main is dependent on its hydraulic gradient (*see* Fig. 10.1), which in turn is set by the static and friction heads; these determine the characteristics of the pumping plant. Air valves are placed at high places and washouts at low points.

With sewage, the length of time between discharge to the wet well of the pumping station and the ultimate discharge at the end of the rising main will be important, because if this time is excessive, the sewage may become septic. The importance of septicity will depend on the amount of sewage being pumped and the age of the sewage when it arrives at the pumping station. Any delay of more than about 12 h will normally warrant further investigation, and probably a revision of the proposals; delays must be calculated on *existing* flows and not on the increased flows likely in the future.

Some years ago, most rising mains were either of cast or spun iron. Many other materials are now available and used regularly; these include steel (usually protected inside and out with bitumen), asbestos-cement, ductile iron and plastics. For larger diameter

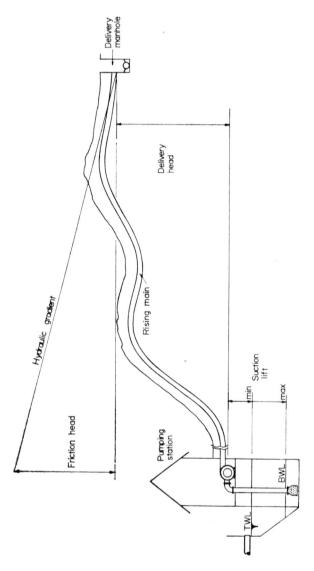

Fig. 10.1. *Hydraulic gradient.*

10 PUMPING MAINS

pipelines, reinforced or prestressed concrete pipes may be used. The pipes to be used must be capable of withstanding the total manometric head on the pumps (static head plus friction losses), as this is the head which will apply in the vicinity of the pumping station. An additional allowance may have to be made for the effects of surge in cases of high pumping velocity, or in very long rising mains. Should the pipeline cross a valley at a lower level than the pumping station itself, the static head at the crossing may be greater.

GENERAL BASIS OF DESIGN

For water supply schemes incorporating river intakes or boreholes, the pumping stations will normally be designed to discharge steadily throughout a large part of the day; this may be over a period of 20 or 22 h. Pumping outputs will therefore be chosen to be slightly in excess of the average 24 h requirements. The outputs of booster stations will, of course, depend on local circumstances, such as storage facilities.

For sewage pumping, the pumps must be capable of pumping the peak flows as they occur in the sewers. These vary throughout the year, the week, and the day. Daily variations range from almost nil at night to something of the order of six times the dry weather flow (d.w.f.) at times during the day. The basis of the figure used for dry weather flow was discussed in Chapter 4.

When sewers are designed to be completely 'separate' (no surface water), the normal maximum flow in a gravity sewer is about 4 d.w.f., and this figure may be used as the maximum to be pumped. Some engineers, however, still prefer to install pumping capacity up to 6 d.w.f. to allow for the effects of some possible acceptance of surface water into the sewers. For a small pumping station only two pumps may be installed, each capable of pumping 4 d.w.f. Where three pumps are installed, each may be capable of about $2\frac{1}{2}$ d.w.f., making about 4 d.w.f. when two pumps are discharging together.

VELOCITY OF FLOW

As pumping for water supply is generally at a fairly constant rate throughout the day, the pumping main is normally designed to give a velocity of about 0·9 or 1·0 m/sec. When pumping will be for short periods only, this rate may be doubled. In a paper by Smith in 1946 [17], it was reported that in Victoria, Australia, a velocity of 0·6 m/sec in all sizes of pipe gave the economical diameter.

For foul sewage pumping it is generally accepted that, taking into account the capital cost of the pumping equipment and the pipeline, and the annual costs of pumping, the most economical diameter of rising main has a velocity of flow at normal pumping rates of between 0·75 and 1·20 m/sec, and a velocity at maximum rates of pumping of not more than about 1·8 m/sec. Some engineers adopt velocities up to 3 or 4 m/sec. The minimum internal diameter suitable for use with crude, unscreened, sewage is 100 mm. Higher velocities may be more economical for *storm* sewage, as the pumps will not be in regular use.

The *approximate* velocity of flow in a rising main is:

$$v = \frac{353 \cdot 4 \, Q}{d^2}$$

Formula 10.1

where

v is the velocity, in m/sec,
Q is the pump discharge, in m^3/h,
d is the internal diameter of the pipeline, in mm.

The velocity in a sewage rising main must be sufficient at some period each day to scour the line and to clear all solid matter. On the other hand, velocities are generally restricted as far as possible to avoid erosion of the pipeline material by grit and stones, and also to minimise friction losses. Where the daily variation in pumping rates is very large, it may be more satisfactory to lay twin rising mains to restrict the maximum velocity to about 2 m/sec. This is referred to in more detail later.

For sludge pumping, the critical minimum velocity is about 1·2 m/sec. Above this the flow is 'turbulent' and the friction head can be based on formulae and tables for liquids (such as Crimp and Bruges), while below this the friction will be much greater than for water. Some workers have suggested friction coefficients of 45 or 50 in the Hazen-Williams formula, as against 100 or more normally used for water. A summary of generally accepted multiplication factors for head loss according to the moisture content of the sludge is shown graphically in Fig. 10.2.

STATIC HEAD AND FRICTION LOSSES

The head on the pump will consist of three elements:

(i) the suction lift,
(ii) the delivery head,
(iii) friction losses.

10 PUMPING MAINS

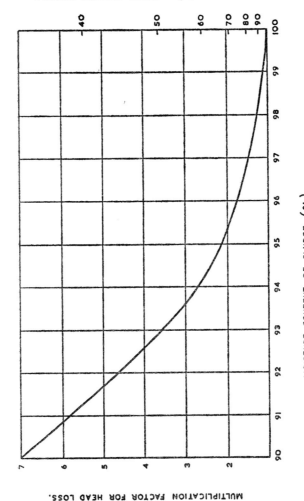

Fig. 10.2. *Friction coefficients and head loss in sludge pipelines.*

The suction lift is the difference in elevation between the source of the supply and the centre-line of the pump. The delivery head is the head above the centre-line to the point of free discharge or to the level of the free surface of the discharge water (*see* Fig. 10.1). Friction losses occur in pipelines, valves, etc. and are dependent on the pipe, etc., material and diameter, the velocity of flow of the liquid, and, to some extent, on the viscosity of the liquid. In this context, the friction head has been taken to include the friction losses through the pump itself and also the velocity head at the point of discharge. This last item can frequently be ignored as it is comparatively small in relation to the other figures. The effects of the variation in head with flow, due to friction losses are shown in Fig. 2.10, where a pipeline friction curve has been plotted in conjunction with various pump characteristic curves.

The suction lift will normally be negative and will therefore *increase* the total head. Occasionally the level of the liquid in the suction pump will be *above* the centre-line of the pump, so that there is a *positive* suction head. This must then be deducted from the delivery head to give the total static head. This is only the case if the elevated sump is connected directly to the pump by the suction pipe; it will not be so if the liquid is fed by gravity into a second suction well at the pumping station itself. The suction lift will vary with variations in the level in the sump; it is important to remember that for sewerage schemes, this level will be at its *maximum* when maximum output is required. This may be important when the suction lift is an appreciable part of the total head, as this increase in level may result in a significant reduction in total head, with a consequent increase in pump discharge.

The delivery head is normally the difference in level between the discharge pipe and the centre-line of the pump. This will not always be so. If there is an intermediate high point in the rising main and if this rises above the hydraulic gradient when based on the point of discharge (*see* Fig. 10.1), then this level will be critical and a revised calculation will be necessary. The latter part of such a pipeline may, in fact, be a gravity line.

The friction head in a pipeline can be determined from tables or formulae. It will usually be necessary to choose a number of possible diameters of pipeline and to calculate the friction losses for each one. Two or three diameters of pipelines will probably then be satisfactory (with the total discharge head within the characteristics of the pumps to be used), and then the choice will be one of economics (see below).

10 PUMPING MAINS

The most suitable total head for an unchokable centrifugal pump for use with sewage is generally of the order of 18 to 21 m.

The friction head in rising mains should be calculated as accurately as possible. It may sometimes be convenient to calculate this head for new pipes and again for pipes after they have been in use for a few years. Most pump manufacturers base the performance of their pumps on the friction losses in *new* pipes. The Manning formula, expressed in metric terms, gives the friction head as follows:

$$h = 7952 \times 10^9 \left(\frac{Qn}{d^{8/3}}\right)^2 \quad \text{Formula 10.2}$$

where

h is the friction head, in m/1000 m,
Q is the flow, in m³/h,
d is the inside diameter of the main, in mm,
n is a friction coefficient (*see* Table 10.1).

TABLE 10.1
FRICTION COEFFICIENTS FOR USE WITH MANNING'S FORMULA

Pipe material	Condition	Value of n
Coated iron and steel pipes	New	0·010
Uncoated iron pipes	New	0·011
Galvanised iron pipes	New	0·012
Vitrified clay pipes	After some weeks of service	0·012
Iron and steel pipes	After use	0·013
All pipes	With imperfect joints and in bad condition	0·015

Other formulae which are frequently used are included in detail in the author's book on sewerage [25].

Head losses in valves and fittings can be calculated from the formula:

$$H = \frac{Cv^2}{2g} \quad \text{Formula 10.3}$$

where

H is the head loss through the fitting, in m,
v is the velocity, in m/sec,
g is the acceleration due to gravity (9·806 m/sec²),
C is a coefficient (*see* Table 10.2).

If the maximum velocity in a rising main is 3·0 m/sec, the velocity head in the pipeline is then:

$$\frac{v^2}{2g} = \frac{3^2}{2 \times 9 \cdot 806} = 0 \cdot 46 \text{ m}$$

and the combined losses due to fittings and valves (*see* Table 10.2) may only be about 1·0 m. These head losses are therefore normally

TABLE 10.2
VELOCITY HEAD IN FITTINGS AND VALVES

Fitting	Coefficient (C)
Bellmouth	0·10
Bend	0·25 to 0·50
Tee-junction	1·00
Sluice valve	0·15
Reflux valve (single flap)	0·70 to 2·00[a]

[a] The value of C for reflux valves increases as the velocity *decreases*.

ignored unless the total manometric head is very low. The total head can be taken as the static head, plus friction losses as calculated earlier, plus something between 1·0 and 1·5 m to account for 'station losses'.

HYDRAULIC GRADIENT

In a gravity pipeline, the hydraulic gradient will be the surface slope of the liquid in the pipeline; this is often assumed to be parallel to the invert. In a pumping main (*i.e.* in a pipeline under pressure) it is the line to which the liquid would rise due to the pressure, if it was not restrained by the pipeline itself. The hydraulic gradient in a pumping main is therefore a line sloping from a maximum at the pump outlet to a minimum at the point of discharge. If the velocity head at discharge is ignored, the hydraulic gradient line will have a value at the pump outlet equal to the sum of the static and friction heads, and it will coincide with the level of the outlet at the point of discharge. This is illustrated in Fig. 10.1.

ECONOMIC DIAMETER

The diameter of a rising main should be selected so that it gives the lowest overall *annual* cost; this cost includes running and maintenance

10 PUMPING MAINS

costs, in addition to the annual repayments of the capital cost, and should include figures for labour and supervision, materials used during maintenance works, and rate charges.

In some cases this selection of a pipe diameter will be a comparatively simple, but time consuming, exercise comparing the cost of one diameter with one or more other diameters. In sewerage design it may also involve the comparison with a possible (but often expensive) gravity sewer laid in a deep trench or tunnel; or it may involve the comparison of the cost of a length of rising main with the cost of a pipeline, part of which operates as a gravity sewer. In all cases, of course, the annual costs used for any comparison must be the total of the annual costs for the pipeline *plus* those for any pumping station. It may be argued that for any particular scheme the overall capital cost of the pumping station itself is not affected to any great extent by an increase or decrease in the friction head due to a change in pipeline diameter, but it is general to take all costs into account for any comparison.

For small pumping stations dealing only with sewage from a separate system of sewers, the size of the rising main will be dictated by the minimum size to avoid blockage and the minimum velocity to prevent silting. These aspects were considered in Chapters 4 and 5, when it was seen that a minimum output of about $17 \text{ m}^3/\text{h}$ is necessary to produce a self-cleansing velocity of 0·6 m/sec in a 100 mm diameter rising main. For these smaller stations, calculations are therefore only required to establish the total head on the pumps.

The economic size of a pumping main depends on the velocity at the *normal* rate of pumping, and not at the peak rate. For sewerage schemes, alternative sizes must be considered which would produce minimum and maximum velocities within the limits set out in Chapter 4. Having obtained the friction head for normal rates of pumping through various alternative diameters, the power requirements can be calculated and the annual cost of pumping against each friction head can be estimated. To obtain the most economical size of rising main, alternative calculations must be made and the capital and annual costs of each compared.

These calculations are a matter of trial and error. A possible size of main is selected, the hydraulic gradient is calculated, and the power requirements are then determined (based on normal rates of pumping). The annual cost of electricity can then be estimated. To this must be added the annual running costs and loan charges of the pumping station. Having assessed the diameter of the pipeline, it is also possible to calculate its own capital cost and then the annual costs, so that a total overall annual cost can be established. A further suitable diameter can then be chosen (to give velocities within the limits agreed) and the total annual costs re-calculated. The various figures obtained can be plotted on a graph for comparison, so that the most economical diameter can be chosen.

DUAL PIPELINES

When the range of pumping duties is too wide to be able to keep the velocities in the pipeline within the limits suggested above, it may be advisable to use dual pipelines. These conditions are more likely to occur in sewerage work than in other fields. A duplicate main is sometimes also used to provide a standby pipeline so that at least one pipe is always available for use.

Where storm sewage is to be pumped in addition to normal flows it may be convenient to arrange for one set of pumps (the d.w.f. pumps) to discharge through one pipeline and for the storm sewage pump to use a separate pipeline. This arrangement allows a reduction in the pump motor power and also provides physical separation of crude and storm sewage. If screens are provided at the overflow to the storm sewage pump well, the storm pumps themselves can be of a higher efficiency.

If there is no particular necessity to separate the flows in the two pipelines, it will be convenient to arrange for connections at the pumping station so that any or all of the pumps can be connected to one or both of the pipelines. If a common delivery pipe inside the pumping station is connected to twin rising mains, then each rising main must have its own sluice valve and washout arrangement.

SURGE PRESSURES

In addition to pressures caused by static and friction heads, allowance must sometimes be made for surge pressure; this is occasionally referred to as *inertia head* and, quite frequently, *water hammer*. This becomes particularly important when the rising main is long (several kilometres) or when a large proportion of the pumping head is friction head.

When a moving column of liquid is accelerated or retarded for any reason, the pressure is changed momentarily so that it rises above, or falls below the value that corresponds to steady flow conditions. This

10 PUMPING MAINS

is more important when the acceleration or deceleration is sudden, and consequently the most common cause of surge pressure or water hammer in a modern pumping installation is the inadvertent stoppage of the prime mover. Some reduction in normal surge pressures can be obtained by arranging the controls so that only one pumping unit will stop or start at a time, but if the pumping station comprises a number of pumps which can operate together, the effect of a sudden failure in the power supply may be quite serious, and in some instances can amount to an increase in head of 50% or more above the working head.

When the retardation rate is a, and the length of the pipeline is l, the *inertia head* is then:

$$h_i = \frac{la}{g} \qquad \textbf{Formula 10.4}$$

where

g is the acceleration due to gravity.

Calculations are however complicated by the 'cushioning' effects of (i) the elasticity or compressibility of the liquid being pumped, and (ii) the modulus of elasticity of the wall of the pipeline. The length of time (in seconds) for any fluctuation in pressure to travel along a pipeline is given by the formula:

$$t = \frac{4l}{v_0} \qquad \textbf{Formula 10.5}$$

where

l is the length, in m,
v_0 is the velocity of the fluctuation, in m/sec.

This velocity will depend on the liquid and on the details of the pipeline (material and wall thickness) but it will normally be of the order of 1220 m/sec.

The theory of the graphical analysis of pressure surges in pumping systems was examined in detail by Lupton in a paper in 1953 [12]. That paper and others are referred to in the *Manual of British Water Engineering Practice* [32] in relation to the construction of Schnyder diagrams. In 1968, Fox published a paper [6] on the use of the digital computer, which he proposed as a more satisfactory alternative to the Schnyder Bergeron graphical method.

Various devices that are used to control surge pressures include by-passes and relief valves, feeder tanks, surge towers and air vessels. Possibly the most effective anti-surge devices are the air vessel (normally sited just outside the pumping station), and feeder tanks located at the points along the pipeline where separation might be expected to occur. A 'Symposium on Water Hammer' was held by the American Society of Mechanical Engineers in 1933 and a paper presented at that Symposium by Enger [5] (reprinted in 1949) put forward a simple energy equation method for finding the size of the air vessel required.

Some relief from surge pressures may be obtained by installing a by-pass from the suction to the delivery at the pumping station, or by a connection, with a reflux valve, at the delivery end of the pipeline, to allow water to feed back when the pressure drops. This latter method is, in effect, a feeder tank at the top of the pipeline.

A paper discussing the effects of valve operation on the surge in pipelines was published by Livingston and Wilson in 1965 [10]. A fast-closing reflux valve will give maximum protection to the pump itself; this can be achieved either by the angle of the gate in the valve, or by designing the valve to make use of the water flow to close the valve quickly. Details of a special plug-type delivery valve suitable for regulating pressure surges were published in 1955 in a paper on the Ashford Common Works of the Metropolitan Water Board [21].

PIPES

Iron pipes may be either cast, spun or ductile. Cast iron pipes with flanged joints are normally used inside the pumping station building. Beyond the pumping station, the pipeline may be of spun or ductile iron, steel, asbestos-cement, or plastics. For large diameter pipelines, prestressed concrete may be economical. The choice of the type of material will be, to some extent, a matter of economics. Other considerations will include the possibility of attacks on the material from either the liquid to be carried or the ground water; flexibility; strength; lightness (for construction in difficult country); and availability of materials (particularly overseas).

Cast and spun iron pipes are available in four classes, 1, 2, 3, and 4. The maximum field hydrostatic test pressures for these classes are set out in Table 10.3. Similar test pressures for standard thickness ductile iron pipelines are set out in Table 10.4.

Steel pipe hydraulic test pressures vary according to the grade of

10 PUMPING MAINS   119

TABLE 10.3
MAXIMUM FIELD HYDROSTATIC TEST PRESSURES FOR GREY IRON PIPELINES

Nominal bore (mm)	Type and method of manufacture	Class	Maximum field hydrostatic test pressure (N/mm^2)
80 to 7000	Spigot and socket centrifugally cast	1 2 3	1·6 2·0 2·5
80 to 300 350 to 600	Flanged centrifugally cast	3 3	2·0 1·6
80 to 700 80 to 300	Flanged sand cast	3 4	1·6 2·0

Notes: 1. Fittings incorporated in a pipeline should have pressure ratings equivalent to the pressure ratings for the pipes.
2. $1 \text{ N/mm}^2 = 10$ bar.

This extract from CP 2010: Part 3: 1972 is reproduced by permission of the British Standards Institution, 2 Park Street, London, W1Y 2BS, from whom copies of the complete publication may be obtained.

TABLE 10.4
MAXIMUM FIELD HYDROSTATIC TEST PRESSURES FOR STANDARD THICKNESS DUCTILE IRON PIPELINES WITH FLEXIBLE JOINTS

Nominal bore (mm)	Maximum field hydrostatic test pressure (N/mm^2)
Up to 300	4·5
350 to 600	3·0
700 to 1200	2·1

Notes: 1. The above pressures are 0·5 N/mm^2 higher than the pressure ratings for the pipes and fittings themselves.
2. $1 \text{ N/mm}^2 = 10$ bar.

This extract from CP 2010: Part 3: 1972 is reproduced by permission of the British Standards Institution, 2 Park Street, London, W1Y 2BS, from whom copies of the complete publication may be obtained.

steel (from grade 22 to grade 27) from 70 kgf/cm^2 (700 m head) for the smaller diameters, to 20 kgf/cm^2 for the larger diameters. Seamless steel pipes are manufactured in two classifications up to 450 mm diameter. Welded steel pipes are available for the larger sizes. Pipes are available with bitumen coating, bitumen lining, or with plastic lining. Jointing of pipes can be by one of a number of welded types; with loose couplings such as Viking Johnson and Victaulic joints; or by patent spigot and socket joint incorporating a rubber ring. Screwed joints are available for the smaller diameter steel tubes (up to 150 mm diameter).

Flanged joints for steel or iron pipes are available for nominal pressures from 2·5 bar to 40 bar.

Asbestos-cement pressure pipes are available in diameters up to 900 mm. Three classes of pipe are manufactured (B, C and D) for working heads from 61 to 122 m. Joints for AC pipes include the AC sleeve 'Widnes' joint and the detachable CI joint; both types incorporate rubber joint rings. Viking Johnson rolled steel couplings are also available for the 300 mm to 900 mm sizes.

Plastics pipes suitable for use as rising mains include PVC and polypropylene. PVC pipes are manufactured in classes B to E and are suitable for heads from 60 to 150 m. Jointing can be with mechanical joints or by solvent welding. Polypropylene pipes are also available in the very small sizes (up to 50 mm); the maximum working head of the largest diameter is 45 m.

Prestressed concrete pipes are produced by a few manufacturers. These are available in the larger diameters only (700 to 1200 mm) and for various proof test heads up to 180 m.

VALVES ON PIPELINES

Sluice valves and reflux valves are fitted to each pump outlet, and further valves are often not necessary unless it is intended to be able to isolate the rising main for washing-out purposes or for the addition of further pumping units in the station at some future date. If so, it is recommended that a sluice valve and reflux valve be installed in a chamber immediately outside the pumping station, in addition to the valves on the pumping units themselves. If two rising mains are installed, an emergency by-pass should be provided between the two pipelines, and sufficient sluice valves must then be fixed immediately outside the station, so that either main can operate as the duty main or both pipelines can operate either separately or together. Long

10 PUMPING MAINS

twin rising mains should be cross-connected at intervals, the connections being fitted with sluice valves.

Where a liquid being pumped will contain solids (*e.g.* sewage), sluice valves should be of the special type suitable for that duty; they should be complete with access doors or plugs to facilitate cleaning. Smaller diameter valves are available in PVC and polypropylene. Butterfly valves and plug valves are used in some circumstances in the water industry. Reflux valves should close quickly to avoid surge pressures; late closure may be due to their design or to the valves sticking due to solid matter jamming the door. The paper by Livingston and Wilson [10] discusses the effects of valve operation on surge pressures in pipelines.

Ball and float valves are used to maintain a constant head in a reservoir or tank; the increased pressure in a pipeline (as the valve closes) can be utilised to control pumps. Foot valves and strainers may be used on suction pipes of pumps where the latter are not self-priming. Table 10.5 gives maximum rates of flow through various types of valve.

As the pipeline will normally be laid to follow the ground contours (minimum cover generally about 1·0 m), provision must be made for the release of trapped air at high points, and for washing out at low points. Air valves must be provided at all high points on the line, and may also be required at intermediate positions along long lengths of even gradient; when used on a sewage rising main these valves must be of the special type suitable for that duty. Air valves are unsuitable for use on *sludge* pumping mains and some engineers do not consider them satisfactory for sewage; in these circumstances vent pipes may be possible but they must be checked to ensure that their outlets will be above the hydraulic gradient line under all conditions.

A washout should take the form of a tee junction and valve; on a water pipeline this can discharge to the nearest watercourse but for a sewage rising main this must be taken to the nearest gravity sewer whenever possible. Where no sewer is available, the washout may have to discharge to a specially constructed sump, which must then be emptied after use. Washout valves and pipelines should be at least 80 mm diameter. Where a section of rising main is arranged to discharge to the wet well of a pumping station, that well must be of adequate capacity, or alternatively, if a second rising main is installed, the pumps can be used to control the level in the well during the period of washing out. It may be convenient to divide a long main

TABLE 10.5
RECOMMENDED MAXIMUM RATES OF FLOW THROUGH VALVES
(litres/sec)

	Size of valve (mm)									
	50	65	80	100	125	150	175	200	250	300
Sluice valve	5·5	10·0	15·0	25·0	40·0	60·0	80·0	100	160	220
Reflux valve	3·0	5·0	8·0	15·0	25·0	37·5	50·0	70	110	160
Foot valve, with strainer	2·2	4·0	6·0	12·0	20·0	30·0	40·0	55	90	130

With acknowledgements to 'Data for Pump Users', published by Sigmund Pulsometer Pumps Ltd and SPP Systems Ltd, Reading, Berks.

10 PUMPING MAINS

into sections by sluice valves so that the capacity of each section is not greater than the capacity of the wet well; if this method is adopted, an air inlet will be required at each valve.

Hatchboxes have been installed on rising mains in the past to provide access points at bends, valves, etc, but these are rarely used now. Some engineers install short lengths of pipe with Victaulic or similar joints at 300 or 400 m intervals, in lieu of hatchboxes. These can be built into special brick chambers or they can be covered over and carefully referenced.

CONSTRUCTION DETAILS

All pipelines which operate under pressure must be adequately anchored at bends, valves, tees, etc. Thrust blocks are usually constructed of *in situ* concrete, and their size will depend on the bearing capacity of the soil at the sides of the trench. The blocks should extend the full width of the trench. Where the hydraulic thrust is in an upward direction, anchor blocks of sufficient weight should be formed to which the pipes can be secured with steel straps. Manufacturers of PVC pipes recommend the use of a thin membrane of bituminised paper, roofing felt or polythene film between the concrete and the pipe; this membrane need only be about 2 or 3 mm thick.

A formula frequently used to calculate the total dynamic and static thrust at any change in direction of a pipeline is:

$$R = 0.002 (H + 0.102 v^2) A \sin \tfrac{1}{2} \varphi \qquad \textbf{Formula 10.6}$$

where

R is the total thrust, in kgf,
H is the total hydrostatic head, in m,
v is the velocity, in m/sec,
A is the cross-sectional area of the pipeline, in mm^2,
φ is the angle of deviation, in degrees.

CHAPTER 11

Specification and Testing

An invitation to a firm of manufacturers to tender for the supply and erection of pumping plant will include a Specification and a Schedule of work to be carried out, along with the usual Conditions of Contract, Form of Agreement, etc. The Specification should include reference to the tests to be carried out both at the works and after delivery to the site.

THE SPECIFICATION
The Specification for a pumping installation should provide a complete picture of the plant required, so that tenders will be submitted on a competitive basis and can be compared one with another. The Specification must fully cover the type of plant and its duty, the quality and finish required, together with details of all auxiliary equipment to be provided. Some scope must, however, be left to the manufacturer to include in his tender the type of equipment which he can supply at the most economical price. A Schedule of some information which should be included in the Specification is given in Table 11.1. Full information should be included regarding the site conditions and access arrangements, the limits of the proposed contract, and the types of test proposed.

Where national standards are to be complied with, these should be referred to in the Specification. Care is required in quoting from national standards, however, and where one standard covers a number of qualities or sizes, the quality or type required must be sufficiently described in the Specification. It is general either to refer to 'the latest standard' in all cases (*i.e.* the latest at the time of tendering), or to quote the actual date of issue of each standard. It is bad practice to mix the two methods in one Specification, as this can lead to confusion during the contract.

While a local authority committee will normally wish to accept the lowest tender in terms of the capital cost, this may not necessarily be

TABLE 11.1

SOME INFORMATION TO BE INCLUDED IN A SPECIFICATION FOR PUMPING PLANT

1. Number of units required
2. Whether for continuous or intermittent operation
3. Capacity:
    (a) individual
    (b) total
4. Head conditions:
    (a) static lift
    (b) suction lift or positive head
    (c) diameter and length of rising main
    (d) number and size, etc. of bends, valves, etc.
5. Nature of liquid to be pumped:
    (a) temperature
    (b) viscosity
    (c) if abrasive or corrosive
    (d) maximum size of solids
6. If to be self-priming
7. Pump type:
    (a) fixed or portable
    (b) vertical or horizontal spindle
    (c) dry well, or wet well
    (d) shafting
    (e) diameter of borehole (if applicable)
8. Details of prime movers and drives
9. Details of starting arrangements
10. If control is to be automatic
11. Any special conditions regarding materials and finish
12. Any special site conditions
13. Details of tests, etc.
14. Delivery date required

the most economical in the long run. Sufficient details should be obtained (in the form of a Schedule) when inviting tenders so that the annual *running costs* of the plant can be compared. These figures should be considered, along with the annual loan charges on the capital outlay, to provide a true comparison of the costs of various offers.

TESTING

An HMSO publication in 1967 [33] reviewed the various testing codes and methods in the USA and the UK, at the time when an attempt was being made to get uniformity in the testing of centrifugal pumps; a section on model testing was included to cater for the increasing size of centrifugal pumps.

Tests of pumping equipment can be divided into three types:

(i) works tests,
(ii) acceptance tests on site,
(iii) routine performance tests.

It should be clear in the Specification, what tests are to be carried out before the plant will be accepted.

EQUIPMENT

The equipment normally used in conjunction with these tests will include the following:

(i) Discharge rate	—measuring tanks
	measuring weirs
	flow meters
	current meters
(ii) Head or pressure	—pressure and vacuum gauges
	mercury or water columns or U-tubes
(iii) Speed of rotation	—tachometer
	rev. counter (and stop watch)
(iv) Power consumption	—electrical readings
	torsion dynamometer

The rates of discharge over a 90° V-notch weir are set out in Table 11.2.

WORKS TESTS

Tests will be carried out at the manufacturer's works, based on the relevant standards, to ensure that the pumps, motors and starters are all in accordance with the Specification. As the purchaser is interested in the ultimate performance of the *complete* equipment, it is preferable if the works test is carried out on the pumping plant complete with all its associated equipment. The pumps are run at

TABLE 11.2

RATES OF DISCHARGE OVER A $90°$ V-NOTCH WEIR
$$Q = 1\cdot 4\, H^{2\cdot 5}$$

Height of water surface above V-notch H (m)	Discharge Q (cumec)
0·05	0·0008
0·06	0·0012
0·07	0·0018
0·08	0·0025
0·09	0·0034
0·10	0·0044
0·12	0·0070
0·14	0·0103
0·16	0·0143
0·18	0·0192
0·20	0·0250
0·22	0·0318
0·24	0·0395
0·26	0·0483
0·28	0·0581
0·30	0·0690
0·32	0·0811
0·34	0·0944
0·36	0·1088
0·38	0·1247
0·40	0·1417
0·42	0·1600
0·44	0·1799
0·46	0·2009
0·48	0·2235
0·50	0·2476

their rated speeds and the head conditions (during the test) are adjusted by opening or closing the discharge valve.

It is usual for the purchaser to specify that his engineer shall have free access to the manufacturer's works at all reasonable times to witness tests. At these tests, special note should be made of any excessive vibration or noise; of any overheating at bearings or glands; and of any leakage from glands, etc. The whole of the equipment should be inspected (inside and outside) as regards standards of finish.

One of the main purposes of the works tests is to ensure that the pumping plant will deliver the specified quantity of water (or sewage, etc.) against the relevant head at the specified speed, and that the power requirements and efficiency are in accordance with the manufacturer's tender. The works test should, however, preferably cover the whole range of the pumping plant's duties. The characteristic curves for a pump are built up from the works performance tests, and are plotted on squared paper as the tests progress. A typical set of these curves was used to prepare Fig. 2.10.

ACCEPTANCE TESTS ON SITE

While it is usual to specify a site test for acceptance of the plant, it may not be possible to enforce a test of output at this stage; this particularly applies to sewage pumps, as the physical characteristics of sewage differ throughout the day and from place to place. A site test should, however, be included to ensure that the plant has been properly installed and correctly aligned so that it runs free from overheating or undue vibration. All bearings, stuffing boxes, and internal running clearances should be properly checked. A site acceptance test must, of course, be carried out in the presence of a responsible representative of the manufacturer.

During the site tests, checks should be made on the quiet and efficient operation of all reflux valves and on the operation of all auxiliary equipment such as electrodes, relays, etc.

ROUTINE PERFORMANCE TESTS

After the original acceptance tests, the pumping equipment should be re-tested from time to time to ensure that its output and efficiency have not deteriorated unduly. It is suggested that after every 5000 h running a performance test should be carried out, along with a physical inspection of the machinery for wear and other damage. These routine tests will show whether the equipment needs repair or replacement due to reduced efficiency, *i.e.* increased power consumption. 'Preventative maintenance' is referred to in Chapter 12.

CHAPTER 12

Operation and Maintenance

While operation and maintenance are rarely the direct concern of the design engineer, it will be apparent that the standards of design of a pumping station will have considerable bearing on the efficiency of the operation and maintenance of the plant. This chapter is included at the end of the book in the hopes that a small insight into the operator's problems might help towards a high standard of design. Standardisation of both the manufacture and the size of pumping units within an area would allow more interchangeability of spares and (where appropriate) of complete pumping units. Such interchangeability would increase general efficiency and decrease the costs of maintenance.

All mechanical and electrical equipment should be kept in a safe condition and good working order at all times. If staff are not available within the organisation, an arrangement should be made for maintenance and repairs to be carried out by a local engineering firm or even by a garage. Many manufacturers are prepared to enter into service agreements for their plant so that it will be inspected and overhauled at regular intervals; where this is arranged, this is an even stronger argument in favour of standardisation.

Preventative maintenance is always preferable to emergency repairs; routine inspections should include the impellers and balancing discs; the bearings; and the general alignment of the shafting. In the split casing type of centrifugal pump, the whole of the rotating elements of the pump are exposed by lifting off the cover. Failure of pumping plant, particularly sewage pumps, may not necessarily be due to any mechanical fault—frequently the trouble will be due to a pump or pipeline blockage (rags, etc.); occasionally it may be loss of prime due to ingress of air.

Records should be kept of all inspections, servicing, repairs and replacements, and all manufacturer's instruction manuals or sheets should be carefully kept and should be easily available for reference. Wherever practicable, instruction books, etc. should be kept *at the site* and not locked away in an office.

Many authorities now operate routine inspection and maintenance schemes using mobile gangs so that a number of pumping stations can be automatically controlled and unmanned. Communication between gangs and their headquarters is vital, and should be easily available either by telephone, or preferably, by radio communication.

PRIMING

Efficient priming of a pump and the suction spaces is essential, since an impeller working in air will not draw water. Air relief valves on the top of pump casings (if fitted) must function correctly and in an automatic station, the levels of the pump cut-in and cut-out must be such that the pump is always primed ready for operation. Where footvalves are fitted these must be checked to ensure that there is no possibility of a pump running 'dry'. When there is loss of prime, it may be due to leaks at pipe joints, valves, glands, etc.

BEARINGS AND BALANCE DISCS

Small pumps may have ball or roller bearings, but larger pumps will probably have thrust bearings, either oil or water lubricated. In a high pressure multi-stage pump a hydraulic balance disc may be fitted to take part of the end thrust; this balance disc or its facing may wear sufficiently to allow the impellers to get out of line. Wear on the disc will be minimised by ensuring that little or no sand or silt is taken up by the pump.

The bearings of all pumps, motors and similar rotating machinery must turn freely and must be kept well lubricated with grease or oil. Oil is preferable where the temperature or speed of rotation is high, but otherwise grease is normal. Care must be taken not to *over-lubricate* the bearings of electric motors.

PUMP GLANDS

Where packed glands are used these will require regular maintenance and the packing must be inspected and renewed from time to time. A faulty gland results in dirty conditions in a sewage pumping installation and it can also cause excessive wear at the shaft and the shaft

12 OPERATION AND MAINTENANCE   131

sleeve. Some slight leakage is normally allowed, to provide cooling for the packing gland and to ensure that the gland is not so tight that rapid wear will occur; this leakage should be about one drop per second for small pumps, and greater for larger units. The wear will normally be taken up by a shaft sleeve to avoid wear on the shaft itself. Where mechanical seals are used in lieu of packing, the manufacturer's instructions should be carefully adhered to as regards maintenance.

If the pump is working under a *positive* suction head, the suction gland must either have a mechanical seal or it must have a water-sealed lantern ring adjacent to the pump impeller so that air will not be drawn into the gland, and also to avoid the possibility of gritty water being forced through the stuffing box.

PUMP IMPELLERS
Sewage pumps are liable to blockage between the impeller and the pump casing as rags and other fibrous matter tend to become wrapped around the impeller. If the pump is not of the split-casing type, handholes should be fitted to give access to the impeller so that blockages can be cleared.

Impellers eventually become worn and must be replaced, as the efficiency of the pump decreases as the clearance between the impeller and the volute casing increases beyond the design clearance. Wear will be more rapid in pumps dealing with sewage or other water which may contain appreciable quantities of grit. It is generally more economical to fit a new impeller than to recondition the old one. Many sewage pumps incorporate 'wear rings' which must be replaced from time to time; these help to minimise the wear on the impeller.

RECIPROCATING PUMPS
Reciprocating pumps are generally very robust and will require little attention beyond routine lubrication. Attention will be required when the flexible diaphragms become worn or damaged, or if fibrous or gritty material gets into the valve.

ELECTRIC MOTORS AND SWITCHGEAR
Electric motors and other electrical equipment generally require very little maintenance, provided they are maintained in clean condition, free from oil, dust, moisture, etc. The bearings of motors must be lubricated to prevent them from running hot but, as stressed above,

care must be taken not to over-lubricate, otherwise oil or grease may get into the electrical connections.

Should a motor run at a speed much less than intended, there may be a break or loose electrical connection, particularly in the rotor lead. Trouble may also be caused by dirty slip rings (in that type of motor) giving poor brush contact; this will be apparent from sparking and flashing, and any worn brushes should be replaced.

VALVES AND PIPELINES

Once installed, pipes and valves should require very little attention. At many pumping stations the sluice valves and penstocks will remain permanently open (or closed), and it is a wise precaution to operate their gates fully a few times once every six months or so. The spindles should, of course, be kept lubricated with grease.

The suction pipes must be checked from time to time to ensure that all joints are air-tight, as any ingress of air will affect the performance of the pump. When strainers are fitted, these must be kept clear of debris, and the foot valves must be examined occasionally to ensure that the flap is clear. Suction pipes must be installed so that their lower ends are immersed at all times and so that there are no 'humps' or 'dips' between the sump and the pump. Where the sump is shallow, the suction pipe should be checked to ensure that it is adequately submerged, otherwise vortices may form, allowing air to be drawn into the pump.

The reflux valves on pump deliveries must be checked to ensure easy operation. Trouble is more likely with sewage pumps than with other types of installation as there is then more likelihood of the valves being jammed with silt and rags. A reflux valve which closes too slowly due to silting will be liable to 'slam' due to pressure surges in the pipeline. For sewage pumping, it is wise to install valves with their spindle bearings or trunnions outside the valve casing; many engineers specify that reflux valves in sewage pumping installations shall have external operating levers so that they can be opened manually to clear any solids. External levers on the reflux valves also allow the pumps and suction pipes to be back-flushed from the rising main; a weight can also be attached to the lever to ensure that the valve closes quickly.

Sluice valves should generally be fully open at all times when a pump is running. At an automatic station this should be checked occasionally. On no account should a pump be regulated by closing

12 OPERATION AND MAINTENANCE 133

a valve on the *suction* side as this will encourage cavitation (*see* Chapter 2).

THE WET WELL, OR PUMP SUMP

Pump suctions will often be protected by screens. For water supply pumps drawing from a river intake, these may consist of a coarse screen followed by a fine screen; the latter will probably be of the band, disc, or drum type. Screenings will be washed off with high pressure jets and returned to the river downstream of the intake.

Sewage pumps will draw from a wet well, which may include screens or comminutors to protect the pumps. Screens are generally not suitable for small, isolated stations. The wet well must be inspected regularly, particularly after a storm, and all grit and debris must be removed. Screenings must not be allowed to accumulate at the station as these will cause nuisance from odour.

GENERAL MAINTENANCE AND RECORDS

Pumphouses and dry wells should be kept clean and tidy at all times. Floors should be mopped over to remove mud and dirt but should not be left in a wet condition. Windows should be cleaned regularly and kept in repair, and paintwork renewed as necessary.

Records should be kept of the hours which each pump operates each day, and of the readings of vacuum gauges, ammeters, etc. Any abnormal variations, or any tendency for readings to steadily increase or decrease must be investigated. Log books should be maintained for all records; these should then be summarised at intervals (*e.g.* monthly) to show the total power used; the power usage according to the quantity pumped (*e.g.* kWh per 10^3 m^3); and the overall efficiencies of the pumps and motors.

References and Bibliography

TECHNICAL PAPERS AND ARTICLES

1. Aitken, I. M. E. et al. (1966). Automatic control of pumping installations, *Effl. Wat. Treat. Manual.*
2. Casson, S. M. (1968). Care in the design of pumping stations, *Rural Dist. Rev.*, **74**, 4, 109.
3. Clements, A. J. (1957). Some of the problems of pumping liquids, *Instn. publ. Hlth. Engrs. J.*, **56**, 2, 96.
4. Creasy, L. R. (1970). Economics and engineering organisation, *Proc. Inst. civ. Engrs.*, Suppl.
5. Enger, L. (1933). Relief valves and air chambers, *J. Amer. Soc. Mech. Engrs.* (Reprinted 1949.)
6. Fox, J. A. (1968). The use of the digital computer in the solution of water-hammer problems, *Proc. Inst. civ. Engrs.*, **39**, Jan. 127.
7. Hood, R. (1964). Fundamental design of sewage pumping stations, *J. Inst. Engrs. Aust.*, **36**, Dec. 309.
8. Humphrey, P. R. (1972). The development of booster pumps in Birmingham, *J. Inst. Water Engrs.*, **26**, 2, 83.
9. Keep, G. A. (1959). Some notes on dual-fuel engines and pumps, *J. Proc. Inst. Sew. Purif.*, 1, 74.
10. Livingston, A. C. et al. (1965). Surges in pipelines—effects of valve operation, *Proc. Inst. Mech. Engrs.*
11. Lupton, H. R. (1948–9). Pumping machinery, *Proc. Inst. civ. Engrs.*, **4**, 291.
12. Lupton, H. R. (1953). Graphical analysis of pressure-surge in pumping systems, *J. Inst. Water Engrs.*, **7**, 2, 87.
13. Lupton, H. R. et al. (1956). Automatic operation of waterworks plant, *J. Inst. Water Engrs.*, **10**, 5, 352.

14. Paish, H. P. S. (1971). Temporary and emergency pumping equipment, *Civ. Eng. and P. W. Review.*, London, Oct. p. 1083.
15. Price, A. V. (1945). Pumps for drainage purposes, *J. Inst. San. Engrs.*, **44**, 2, 71.
16. Ross-Smith, A. J. (1972). Rochester's stormwater drainage, *Survr. Local Gov. Tech.*, **140**, 4201, 26.
17. Smith, D. B. (1946). Rising mains—most economic diameter, *Proc. Inst. civ. Engrs.*, **8**, October, 534.
18. Summerton, D. C. (1962). Electrically-operated pumping plant and equipment, *J. Inst. Water Engrs.*, **16**, 1, 13.
19. Wauchope, G. A. et al. (1952). The design of large pumping installations for low and medium heads, *Proc. Inst. Mech. Engrs.*
20. Willis, J. S. M. et al. (1968). Telemetry symposium, *J. Inst. Water Engrs.*, **22**, 3, 170.
21. Wood, F. (1955). Ashford common works, with particular reference to the pumping plant, *J. Inst. Water Engrs.*, **9**, 4, 311.

TEXTBOOKS
22. Addison, H. (1966). *Centrifugal and Other Rotodynamic Pumps*, Chapman and Hall Ltd., London.
23. Allen, E. (1960). *Using Centrifugal Pumps*, Oxford University Press, London.
24. Anderson, H. H. (1962). *Centrifugal Pumps*, Trade and Technical Press Ltd., London.
25. Bartlett, R. E. (1970). *Public Health Engineering: Design in Metric—Sewerage*, Applied Science Publishers, London.
26. Bartlett, R. E. (1971). *Public Health Engineering: Design in Metric—Wastewater Treatment*, Applied Science Publishers, London.
27. Keefer, C. E. (1940). *Sewage-Treatment Works*, McGraw-Hill Book Co. Inc., New York.

MISCELLANEOUS
28. *An Introduction to Engineering Economics* (1969). Inst. civ. Engrs.
29. *Choosing Valves* (1949). Glenfield & Kennedy Ltd.
30. *Data for Pump Users* (1972). Sigmund Pulsometer Pumps Ltd.
31. *Design and Construction of Sanitary and Storm Sewers* (1966). Water Pollution Control Federation Manual of Practice No. 9; Amer. Society of Civil Engrs. Manual of Engng. Practice, No. 97.

32. *Manual of British Water Engineering Practice* (1969). Inst. Water Engrs.
33. *Pump Design, Testing and Operation* (1967). Min. of Technology, HMSO.
34. *Safety in Sewers and at Sewage Works* (1969). Inst. civ. Engrs.
35. *Steel Pipes for Water Mains* (1971). British Steel Corpn., Tubes Division.

APPENDIX A

Definitions and Abbreviations

DEFINITIONS

Catchment Area:	The area of a watershed discharging to a sewer, river or lake, etc.
Combined Sewer:	A sewer designed to carry both foul sewage and surface water.
Crude Sewage:	Sewage which has received no treatment.
Datum:	A plane of reference for a system of levels.
Dry Weather Flow: (d.w.f.)	The daily rate of flow of sewage, together with infiltration (if any), in a sewer in dry weather—measured after a period of seven consecutive days of dry weather during which the rainfall has not exceeded 0·25 mm.
Ejector:	A means of raising sewage by admitting it through a valve to a closed vessel, and then ejecting it through another valve by admitting compressed air into the vessel.
Foul Sewage:	Any water contaminated by domestic waste or trade effluents.
Gradient:	The inclination of the invert of a pipeline expressed as a fall in a given length.
Gravity Sewer:	A sewer in which the sewage runs from one end to the other on a descending gradient, and when pumping is not required.
Hydraulic Gradient:	The free surface slope of a liquid in a pipeline. This is generally taken as parallel to the invert in a gravity sewer.

Invert:	The lowest point of the internal cross-section of a sewer or channel.
Rising Main:	A pipeline carrying the discharge from a pump or pumps, which is running full and at a pressure greater than atmospheric.
Separate Sewer:	A sewer designed to carry foul sewage only, or surface water only.
Sewage:	Water-borne human, domestic and farm waste. It may include trade effluent, subsoil or surface water.
Sewer:	A pipeline to carry sewage and other wastes, and not normally flowing full.
Sewerage:	A system of sewers and ancillary works to convey sewage from its point of origin to a treatment works or other place of disposal.
Sludge:	Accumulated suspended solids deposited in pipes or tanks, mixed with water to form a semi-liquid substance.
Storm Sewage:	Foul sewage diluted with surface water.
Storm Sewage Overflow:	A weir, siphon or other device, on a combined or partially-separate sewerage system, introduced for the purpose of relieving the system of flows in excess of a selected rate, so that the size of the sewers downstream of the overflow can be kept within economical limits, the excess flow being discharged to a convenient watercourse.
Surface Water:	Natural water from the ground surface, paved areas and roofs.

ABBREVIATIONS

Ampere(s)	A
Cubic metre(s)	m^3
Cubic metres per second	m^3/sec (or cumec)
Day(s)	d
Dry weather flow	d.w.f.
Gravitational acceleration (9·806 m/sec^2)	g
Hectare (10^4 m^2)	ha
Hour(s)	h
Hertz (frequency)	Hz

APPENDIX A

Kilogramme(s)	kg
Kilogramme-force	kgf
Kilometre(s)	km
Kilovolt-ampere(s)	kVA
Kilowatt(s)	kW
Metre(s)	m
Metres per second	m/sec
Millimetre(s)	mm
Minute(s)	min
Newton(s)	N
Revolutions per minute	rev/min
Second(s)	sec
Square metre(s)	m^2
Year (annum)	a

APPENDIX B

Conversion of Flow Rates

Litres per second (litres/sec)	Cubic metres per hour (m^3/h)	Cubic metres per hour (m^3/h)	Litres per second (litres/sec)
1	3·6	1	0·277
2	7·2	2	0·555
3	10·8	3	0·833
4	14·4	4	1·111
5	18·0	5	1·388
6	21·6	6	1·666
7	25·2	7	1·944
8	28·8	8	2·222
9	32·4	9	2·500
10	36·0	10	2·777
12	43·2	12	3·333
14	50·4	14	3·888
16	57·6	16	4·444
18	64·8	18	5·000
20	72·0	20	5·555
25	90·0	25	6·944
30	108·0	30	8·333
35	126·0	35	9·722
40	144·0	40	11·111
45	162·0	45	12·500
50	180·0	50	13·888
55	198·0	55	15·277
60	216·0	60	16·666
65	234·0	65	18·055
70	252·0	70	19·444
75	270·0	75	20·833
80	288·0	80	22·222
85	306·0	85	23·611
90	324·0	90	25·000

APPENDIX B

Conversion of Flow Rates—*contd*

95	342·0	95	26·388
100	360·0	100	27·777
110	396·0	110	30·555
120	432·0	120	33·333
130	468·0	130	36·111
140	504·0	140	38·888
150	540·0	150	41·666
160	576·0	160	44·444
170	612·0	170	47·222
180	648·0	180	50·000
190	684·0	190	52·777
200	720·0	200	55·555
250	900·0	250	69·444
300	1 080·0	300	83·333
350	1 260·0	350	97·222
400	1 440·0	400	111·111
450	1 620·0	450	125·000
500	1 800·0	500	138·888
600	2 160·0	600	166·666
700	2 520·0	700	194·444
800	2 880·0	800	222·222
900	3 240·0	900	250·000
1 000	3 600·0	1 000	277·777

Note: 1000 litres/sec = 1 cumec.

APPENDIX C

Conversion Factors

Unit	Imperial		Metric
Length	0·621 4 miles	1	km 1·609
	3·281 ft	1	m 0·304 8
	1·094 yards	1	m 0·914 4
	0·049 7 chains	1	m 20·116 8
	0·547 fathoms	1	m 1·829
	0·039 37 in	1	mm 25·40
	0·039 37 'thou'	1	μm 25·40
Area	$1\cdot550 \times 10^{-3}$ in^2	1	mm^2 645·2
	10·764 ft^2	1	m^2 0·092 90
	1·196 yd^2	1	m^2 0·836 1
	0·386 1 sq miles	1	km^2 2·590
	2·471 acres	1	ha 0·404 7
Volume	35·315 ft^3	1	m^3 0·028 32
	1·308 yd^3	1	m^3 0·764 6
	$0\cdot061 \times 10^{-3}$ in^3	1	mm^3 16 387·1
Capacity	0·220 Imp gal	1	litres 4·546
	0·264 2 US gal	1	litres 3·785
	1·760 pints	1	litres 0·568
Velocity	3·281 ft/sec	1	m/sec 0·304 8
	196·8 ft/min	1	m/sec 0·005 1
	0·621 4 m.p.h.	1	km/h 1·609
	0·539 6 knots (UK)	1	km/h 1·853 2
Mass	0·984 2 ton	1	tonne 1·016
	0·019 7 cwt	1	kg 50·802
	2·205 lb	1	kg 0·453 6
	0·035 3 oz	1	g 28·349 5
Mass/unit area	29·5 oz/yd^2	1	kg/m^2 $33\cdot90 \times 10^{-3}$
	0·001 4 lb/in^2	1	kg/m^2 703
	$0\cdot398 \times 10^{-3}$ ton/acre	1	kg/ha 2 510·71

Conversion Factors—contd

Unit	Imperial		Metric
Rate of flow or discharge	13·20 gal/min	1	litres/sec 0·075 7
	35·31 cusec	1	cumec 0·028 3
	2118·6 cumin	1	cumec $0·47 \times 10^{-3}$
	19·01 m.g.d.	1	cumec 0·052 6
	3·675 gal/min	1	m^3/h 0·272
	219·97 gal/day	1	m^3/day 0·004 5
	14·3 ft^3/1 000 acres	1	litres/ha 0·070
	0·091 5 cusec/sq mile	1	litres/sec km^2 10·933
Density	10·001 lb/gal	1	kg/litre 0·099 8
	0·062 43 lb/ft^3	1	kg/m^3 16·02
Force	0·224 8 lbf	1	N 4·448
	0·100 4 tonf	1	kN 9·964 0
	7·233 pdl	1	N 0·138 3
Force or weight per unit length	0·068 5 lbf/ft	1	N/m 14·593 9
Pressure or stress	$0·145 \times 10^{-3}$ lb/in^2	1	N/m^2 6 894·76
	0·004 01 in H_2O	1	N/m^2 249·089
a {	0·335 ft H_2O	1	kN/m^2 2·989 1
	0·295 3 in Hg	1	kN/m^2 3·386 4
Power or energy	1·341 hp	1	kW 0·745 7
	0·009 5 therm	1	MJ 105·5
	0·277 8 kWh	1	MJ 3·6
	0·947 8 Btu	1	kJ 1·055 1
Miscellaneous:			
temperature	1·8 deg F	1	deg C 0·555
mass/length	0·672 0 lb/ft	1	kg/m 1·488
Bending moment	8·851 lbf in	1	Nm 0·113 0
2nd moment of area	$2·4 \times 10^{-6}$ in^4	1	mm^4 416 231
	116·0 ft^4	1	m^4 0·008 63
Shear strength	20·885 lbf/ft^2	1	kN/m^2 0·047 9
Bearing capacity	0·009 3 ton/ft^2	1	kN/m^2 107·252
Heat flow	3·41 Btu/h	1	W 0·293 1
Thermal conductivity	6·933 $Btu/in/ft^2$ h deg F	1	W/m deg C 0·144 2
Coefficient of heat transfer	0·176 1 Btu/ft^2 h deg F	1	W/m^2 deg C 5·678 3
Calorific value	0·43 Btu/lb	1	kJ/kg 2·326
	0·026 9 Btu/ft^3	1	J/litre 37·259
Hydr. loading	169·5 gal/yd^3 day	1	m^3/m^3 day 0·005 9
Filtration rate	185·5 gal/yd^2 day	1	m^3/m^2 day 0·005 4
	0·87 gal/ft^2 h	1	m^3/m^2 day 1·15

continues

Conversion Factors—*contd*

Unit	Imperial		Metric
Air supply	2·12 ft^3/min	1	litres/sec 0·471 9
	0·164 ft^3/gal	1	m^3/m^3 6·1
Others	183·9 gal/yd^2	1	m^3/m^2 5·437 × 10^{-3}
	67·2 gal/ft day	1	m^3/m day 0·014 9
	20·4 gal/ft^2 day	1	m^3/m^2 day 0·048 9
	2·82 miles/gal	1	km/litre 0·354
Costs	10·9 p 1 000 gal	1	p/m^3 0·092

[a] These items are based on standard conditions when:

1000 mbar (1 bar) = 29·53 inches Hg
1 mbar = 100 N/m^2
1 bar = 100 kN/m^2
1 N/m^2 = 1 Pascal (Pa)

°C = 5/9 (°F − 32)

Index

Abbreviations, 138
a.c. supply, 57, 58
Acceptance tests, 128
Access, 96, 98, 100, 104, 124
Accessories, 64
Activated sludge, 8, 32, 45, 46
Adams 'Autoram', 17
Air
   lift pumping, 15, 24, 30, 61
   release pipes, 84, 97
   valves, 107, 121
   vessels, 11, 13, 118
Alarms, 66, 69, 70, 73, 97
Alternating double filtration, 47
Alternative fuel engines, 60
Amenity, 94
Ammeters, 66
Anaerobic conditions, 35
Ancillaries, 81
Annual costs, 42, 114, 125
Archimedian screw, 6, 7, 9, 27
   (*see also* Screw pumps)
Architecture, 101
Asbestos–cement pipes, 120
Automatic control, 2, 26, 27, 29, 39, 56, 60, 63, 66, 83
Auto-transformer starters, 25, 57, 64
Auxiliary suctions, 49, 86, 100
Axial flow pumps, 6, 18, 23, 26, 27, 46, 79

Balancing discs, 130
Ball valves, 121

Banks, 106
Bearings, 130
Bellmouths, 82
Benching, 98
Bends, 114
Biological filtration, 46
Booster pumps, 23, 28, 58, 71, 103
Borehole pumps, 6, 12, 15, 23, 24, 58, 103
Brickwork, 101
Brownson floatless switch, 72
Buildings, 96
Bushes, 87
By-passes, 120

Cables, 25, 78
Capacities, sewage pumping, 33, 39, 53, 64, 98
Capacitors, 59, 75
Cast iron pipes, 118, 119
Cavitation, 20, 133
Cellar drainage pumps, 49
Centrifugal pumps, 3, 19, 23, 28, 57
Chain-link fencing, 105
Characteristic curves, 20, 87, 112, 128
Circuit breakers, 73
Combined sewers, 33, 42, 96
Comminutors, 41, 86, 133
Commutators, 57, 74
Compressed air
   ejectors—*see under* Ejectors
   power, 61

Computers, 66, 117
Concrete, 98
Condensation, 24
Condensers, 78
Conductivity, 67
Contacts, 87
Contractors' pumps, 12, 55
Controls, 63
Conversion factors, 142
Costs, 42, 59
Cranes, 91
Crimp and Bruges, 110
Cut-outs, 73

Dall tubes, 67
d.c. supply, 57
Definitions, 137
Delivery
  head, 110
  pipes, 82
Design point, 19
Detention time, 35
Diaphragm pumps, 12
Diesel engines, 59
Diffuser chamber, 4, 5
Direct-on-line starting, 25, 57, 63
Disintegrating pumps, 4, 41, 49, 53
Drip-proof motors, 58
Dry weather flow, 32, 109
Dry well, 38, 49, 86, 99
Drying room, 95
Dual-fuel engines, 60
Ductile iron pipes, 118, 119
Duplicate pipelines, 34, 116, 121
Duty pump, 39, 74

Earth faults, 73
Earthing, 78
Economic diameter of rising main, 43, 110, 114
Efficiency, 23, 27, 51, 57
Ejectors, 12, 51
Electric
  motors, 56, 131
  wiring, 78
Electrodes, 47, 68

Emergency
  lighting, 66, 75
  overflow, 32, 96
Erosion, 20

Farms, 33
Feeder tanks, 118
Fencing, 105
Fine screens, 27
Fittings, 113
Float
  control, 47, 67
  valves, 121
Floatless control, 49, 68
Flocs, 8
Flood level, 32, 96
Floor plating, 89, 100
Floors, 41, 49, 89, 98, 100
Flotation, 98
Flow formulae, 110, 113
Flowering shrubs, 106
Foot valves, 85, 121, 122, 132
Freezing, 91, 103
Friction head, 107, 110, 113
Fuel oil storage, 60
Fuses, 87

Gantries, 91
Gas engines, 59
Gates, 105
Gauges, 85
Gear type pumps, 7
Gland
  drainage, 86
  seals, 29, 42, 86, 95, 130
Glands, 42, 49, 130
Glass, 103
  blocks, 104
Graphical analysis, surge pressures, 117
Grassed areas, 106
Gravity sewers, 43
Grey iron pipes, 118, 119
Grit, 15, 35, 41, 42, 86, 104, 133
Guards, 95
Gullies, 105

INDEX

Handholes, 42
Handrailing, 89, 95, 100
Handwheels, 83
Hardy Spicer couplings, 62
Hatchboxes, 123
Hazen–Williams formula, 110
Head, 43, 46
Headstocks, 84
Heaters, 75
Heating and lighting, 66, 91, 103
Heavy duty starters, 64
Helical rotor pumps, 8
High lift pumps, 27
High torque motors, 57
Hoists, 91
Horizontal-spindle pumps, 2, 5, 23, 38, 99
Hours run meters, 66
HRC fuses, 73
Hydraulic
   gradient, 108, 112, 114, 115
   meters, 66
   ram, 17
Hydrogen sulphide, 35

Impellers, 4, 5, 131
Indicating lights, 73, 74
Inductance, 58
Industrial wastes, 32
Inertia head, 116
Infiltration, 32, 33
Inlet chamber, 35, 41, 98
Insulation, 103
Interchangeability, 129
Interest charges, 43
Intermittent duty starters, 64
Internal combustion engines, 56, 59
Irrigation, 26
Isolating switches, 66

Krondorffer connection, 64

Ladders, 87
Lagging power factor—*see under*
   Power factor

Land drainage pumps, 6, 23, 26, 64, 79
Landscaping, 105
Layout of pumping stations, 35
Leading eye, 5, 41
Lift and force
   ejectors—*see under* Ejectors
   pumps, 11
Lifting tackle, 91
Lighting, 91, 98, 100, 103
Loadbearing walls, 101
Loan charges, 43, 115
Low head pumps, 6
Lubricants, 87
Lubrication, 130

Macerator pumps, 31, 42
Magnetic flow meters, 67
Maintenance, 96, 129, 133
Manning's formula, 113
Manometric head, 18, 20, 109, 114
Mechanical gland seals, 29
Mercury switches, 70
Messroom, 95
Metering pumps, 55
Meters, 66
Methane gas, 35
Microstrainers, 47
Mixed flow-pumps, 7, 19, 26, 34, 49
Modules, 103
Mortars, 102
Motors, 56
Multiplexing, 67
Multi-speed pumps, 27, 39, 71, 79
Multi-stage pumps, 5, 25
Mutrators, 54

No-volt release, 66, 73
Non-overloading characteristics, 6, 7, 25
Nuisance, 31

Odours, 31, 34, 35, 51
Office, 95

INDEX

Oil engines, 59, 104
Operation, 129
Ordinary duty starters, 64
Orifice plate, 67
Ornamental trees and shrubs, 106
Output, 4
Overflow, 32, 96
Overload release, 66, 73

Packaged units, 51
Packings, 87
Parking facilities, 95
Penstocks, 35, 67
Performance tests, 126
Petrol engines, 59
pH, 67
Pipe supports, 81
Pipework, 81, 118, 132
Plastics pipes, 120
Plinths, 100
Plug and socket board, 67, 74
Plunger pumps, 12, 46
Pneu pump, 16
Pneumatic
  bubbler, 71
  compression tank, 29
  ejectors—*see under* Ejectors
  switch, 29, 73
Portable pumps, 46, 55
Power
  factor, 44, 58, 75
  failure, 56, 117
  input, 22, 43, 115
Prefabricated structures, 103
Preferred speed, 58
Pressure
  gauges, 85
  operated switch, 29, 73
  variations, 67
Prestressed concrete pipes, 120
Preventative maintenance, 129
Prime movers, 56
Priming, 85, 97, 99, 130
Probes—*see under* Electrodes
Propeller pumps—*see under*
      Axial-flow pumps
Puddle flanges, 82

Pumping mains—*see under* Rising
   mains
Push-button control, 66

Radio control, 67, 95, 130
Rags, 41
Rainfall, 34
Ram pumps, 11, 17
Rattling, 20
Reciprocating pumps, 11, 23, 29, 131
Recirculation, 46, 67
Recorders, 66
Records, 130, 133
Reflux valves, 83, 120, 122, 132
Reinforced concrete, 98
Relief valves, 118
Re-pumping, 29
Rising mains, 43, 51, 53, 107
River intakes, 26, 79
Roads, 104
Rotary pumps, 7
Rotor circuit resistance, 57, 80
Running costs, 42

Safety, 87, 94
Sand filters, 47
Schnyder diagrams, 117
Screenings, 41, 86, 104
Screens, 26, 27, 32, 41, 53, 86, 133
Screw pumps, 7, 8, 27, 46
Sealtrode, 69
Self-cleansing velocity, 31, 34, 40, 51
Self-priming pumps, 84
Separate sewers, 33, 109
Septicity, 32, 34, 35, 39, 107
Sewage
  pumps, 4, 7, 11, 31, 45, 64, 67, 82, 84, 94, 97
  treatment, 1, 15, 32, 45, 55
  liquors, 46
Sewerage, 31
Shafting, 62
Short circuit, 73
Siting, 96

INDEX

Slip-ring motors, 57, 78
Sludge
  sewage, 11, 32, 45, 110, 121
  pipelines, 45
Sluice valves—*see under* Valves
Small installations, 30, 53, 61, 103
Solids, 4, 31, 41
  diverter, 53, 103
Spares, 87
Specific speed, 18, 27
Specification, 87, 124
Speed of rotation, 18, 42, 58
Split casing pumps, 4, 42, 129
Spun iron pipes, 118
Squirrel cage motors, 25, 56
Stairways, 87, 100
Standardisation, 129
Standby units, 39, 51, 60
Star-delta starters, 57, 63
Starters, 25, 26, 57, 63
Starting currents, 57, 63
Station losses, 114
Stator-rotor starters, 57, 63
Steam power, 12, 60
Steel pipes, 118
Storm sewage, 31, 49, 56, 110, 116
  overflows, 33, 44, 96
Strainers, 132
Submersible pumps, 2, 5, 23, 25, 29, 38, 47, 50, 58, 99, 103
Substructure, 99
Suction
  lift, 110
  pipes, 4, 38, 40, 82, 132
Sump pumps, 16, 49, 86, 100
Supernatant liquors, 45, 47
Superstructure, 101
Surcharge of sewers, 34
Surface water, 34, 105 (*see also* Storm sewage)
Surge, 109, 116
Switchgear, 64, 131
Synchronous motors, 57

Telemetry, 63, 67, 74
Telephone, 95, 130
Temperature, 67

Tenders, 125
Testing, 126
Thrust
  bearings, 28, 130
  blocks, 123
Time
  delays, 66, 68
  concentrations, of, 34
Toilet facilities, 35, 94
Tools, 87
Torque, 3, 57, 63
Totally enclosed motors, 58
Transmitters, 66
Trees, 106
Tropical conditions, 104
Turbine pumps, 5
Type of pumps, 3

Unchokable pumps, 4, 34, 41, 47, 53, 113
Unclimbable fencing, 105
Underground stations, 32, 51, 96, 103

Vacuum gauges, 85
Valves, 67, 83, 100, 116, 120, 122, 132
Variable
  pitch impellers, 7, 26, 27, 71, 79
  speed motors, 27, 39, 57, 71, 79
Velocity
  head, 110, 114
  flow, of, 34, 43, 51, 109
  (*see also* Self-cleansing velocity)
Vent pipes, 94, 98
Ventilation, 35, 94, 98, 103
Venturi meters, 67
Vertical spindle pumps, 2, 5, 23, 38, 40, 84, 89, 97, 99
Vibration, 20, 127
Viscosity, 110
V-notch weir, 126, 127
Voltage, 64
Voltmeter, 66

Volute, 4, 131
Vortex, 41, 71, 83, 132

Walls, 100, 101
Washing facilities, 35, 94
Washouts, 107, 121
Washwaters, 47
Water
  consumption, 32
  hammer, 116
  power, 62

Water—*Contd*
  sealed glands, 42
  supply pumps, 5, 6, 7, 8, 23, 109
Weatherproof motors, 103
Wells, 23, 24
Wet well, 34, 35, 38, 46, 52, 82, 85, 97, 133
Wind power, 27, 62
Windows, 94, 103
Wiring, 78
  diagram, 87
Works test, 126